Pharmaceutical Unit Operations

Coating

■ ■ ■

Drug Manufacturing Technology Series

Volume 3

Edited by

Kenneth E. Avis
Atul J. Shukla
Rong-Kun Chang

CRC Press
Taylor & Francis Group
Boca Raton London New York

CRC Press is an imprint of the
Taylor & Francis Group, an **informa** business

CRC Press
Taylor & Francis Group
6000 Broken Sound Parkway NW, Suite 300
Boca Raton, FL 33487-2742

First issued in paperback 2019

ISBN-13: 978-1-57491-082-7 (hbk)
ISBN-13: 978-0-367-40033-0 (pbk)

Library of Congress Cataloging-in-Publication Data

Pharmaceutical unit operations : coating / Kenneth E. Avis, et al., editors.
  p. ; cm - (Drug manufacturing technology series ; v.3)
  Includes bibliographical references and index.
  ISBN-13: 978-1-5749-1082-7 (alk. paper)
  ISBN-10: 1-5749-1082-5 (alk. paper)
  1. Drugs--Coatings. I. Avis, Kenneth E., 1918-
  II. Shukla, Atul.
  III. Chang, Rong-Kun. IV. Series

  RS199.C63P47 1998
  615'.19--dc21   98-27219

Visit the Taylor & Francis Web site at
http://www.taylorandfrancis.com

and the CRC Press Web site at
http://www.crcpress.com

CONTENTS

5.   Automated Coating Systems   223

Gregory Raymond Smith
James G. Spencer

FOREWORD

As stated in the Foreword of the first two volumes in the *Drug Manufacturing Technology Series* of applied reference books, these books are intended to be practical, focusing on how to perform the processes and tasks covered. The series is not planned to be a continuum of closely related topics in pharmaceutical processing; rather, it is a series of topics that may be quite varied but of current interest to practitioners in their respective fields. Consequently, this third volume varies in subject matter from the first two but covers a timely subject, "Pharmaceutical Unit Operations: Coating."

The coating of solid dosage forms of therapeutic agents is both old and new. Sugar coating of tablets has been practiced for a long time, largely as an art. Today, it has become a scientific technology. Even more technologically sophisticated are film coating and microencapsulation processes. Practical aspects of performing all three of these technologically advanced unit operations are the focus of this book. It is anticipated that in the future there will be other books on other types of unit operations, but coating received the nod as being the most needed topic at this time.

Supplemental to the three processing chapters are three, closely related topics presented as additional chapters. First, because automation is so much a part of the three processes discussed, the development of automation to its current degree of complexity was highly important. Therefore, the topic of automation has been developed within this reference volume. Next, adequate and effective cleaning of equipment used in the processes and its validation has become increasingly critical and of regulatory significance. Without addressing this topic, the book would have had a critical omission. And, finally, with all of the sophistication of the processes for manufacturing coated products, their stability and the control of their quality continue to be subjects of extreme importance. Thus, this book has been rounded out to include these three supplemental topics.

Because an extensive index is the key to the utility of a reference book, every effort has been made to provide you with a complete index for this volume. In addition, to enhance the usefulness of closely related volumes in the **Drug Manufacturing Technology Series**–volumes on Sterile Dosage Forms and volumes on Nonsterile Dosage Forms–we have created continuing Key Concept Indexes. *Pharmaceutical Unit Operations: Coating,* the first volume dealing with Nonsterile Dosage Forms, initiates a new Key Concept Index. Since it is our desire to make this series of books not only practical but also highly useful, we hope that you find the continuing Key Concept Indexes valuable tools.

I have been privileged to have had a role in bringing this third volume into being. I hope that you will have as much pleasure and benefit from using it as I have had in seeing it come together in its now completed form.

Kenneth E. Avis, D.Sc.
Coordinating Editor

Other Volumes in **Drug Manufacturing Technology Series**

Volume 1:   *Sterile Pharmaceutical Products:*
Process Engineering Applications
Edited by Kenneth E. Avis

Volume 2:   *Biotechnology and Biopharmaceutical Manufacturing,*
Processing, and Preservation
Edited by Kenneth E. Avis and Vincent L. Wu

AUTHOR BIOGRAPHIES

JOHN F. ADDISON

Mr. Addison graduated from the University of Southern Mississippi with majors in chemistry and biology and received his Master's in Chemistry from Memphis State University. After working at Baxter Laboratories and Velsicol Chemical, he continued his career at Schering-Plough HealthCare Products as Senior Methods Chemist in 1966 and progressed to Vice President, Quality Control and Technical Services until early retirement in 1997. He was active in cGMP regulatory areas with the Nonprescription Drug Manufacturers Association and the FDA and continues to give conference presentations on regulatory matters.

KENNETH E. AVIS

Dr. Avis is a distinguished leader in pharmaceutical science. He is past president of the PDA and has served the U.S. Pharmacopeia and U.S. FDA in numerous advisory capacities. Dr. Avis has consulted with more than 50 pharmaceutical companies, universities, hospitals, and governmental agencies; published over 30 peer-reviewed research papers; and is currently serving as Editor for the *Drug Manufacturing Technology Series* for Interpharm Press while maintaining his affiliation with the University of Tennessee, Memphis, as Emeritus Professor in Pharmaceutics.

EDDIE L. BRUNSON

Mr. Brunson is currently a Principal Scientist, Pharmaceutical Formulations Research & Development at Schering-Plough HealthCare Products. He has 20 years' experience in formulation and technical support of solid dosage products. His experience includes conventional and high shear granulation processing, conventional sugar coating, film coating, and fluid-bed processing.

ROBERT J. CAMPBELL

Mr. Campbell is a Principal Engineering Associate for Chemical Engineering and Instrumentation Consultants, Inc. (CE&IC), in Burlington, New Jersey. His experience includes development, design, validation, and start-up of fluid bed drying and tablet coating systems, as well as state-of-the-art instrumentation and control systems. Before joining CE&IC, he served as Vice President of Engineering at Vector Corporation. He is a member of the Institute of Electrical and Electronics Engineers. He attended the University of Arizona and majored in electrical engineering.

RONG-KUN CHANG

Dr. Chang is currently Senior Principal Scientist in Pharmaceutical Development at Shire Laboratories. He previously held similar positions at DuPont Merck Pharmaceutical Company and Schering-Plough Research, where he was responsible for formulation research and process development for a variety of dosage forms, including sustained release products. He served as a teaching and research assistant at the University of Georgia, School of Pharmacy, while pursuing a Ph.D. degree there. He has 35 publications in pharmaceutical and related journals.

WILLIAM E. HALL

Dr. Hall is an international expert on the subjects of cleaning and process validation. He has 40 years of pharmaceutical experience, including 22 years at GlaxoWellcome in the areas of R&D, quality assurance, and compliance. Prior to that he was professor at the University of North Carolina at Chapel Hill for 7 years. Since 1995, he has served as a consultant to more than 100 companies worldwide and has made technical presentations in the United States, Canada, Mexico, Switzerland, Germany, England, Finland, Austria, Sweden, Norway, and Japan. He has also published extensively in books and journals and serves on the editorial board of the *Journal of Validation Technology.*

IRWIN C. JACOBS

Dr. Jacobs has worked in the field of controlled release and microencapsulation for approximately 15 years. He has been on the faculty of Rockhurst College (Kansas City, Missouri) and Washington University (St. Louis, Missouri). He is currently Director of Chemical Technology at Particle and Coating Technologies, Inc. His interests include the formulation development aspects of spray drying, particle coating methods, and bioavailability enhancement techniques.

NORBERT S. MASON

Dr. Mason has worked on applications of microencapsulation for the food and pharmaceutical industries and has developed processes for forming, coating, and sizing particles and synthetic membranes for medical uses for more than 30 years. He was research professor at Washington University (St. Louis, Missouri) until 1995 when he joined Particle and Coating Technologies, Inc., as Director of Engineering.

GARY L. SACKETT

Mr. Sackett is Laboratory Manager at Vector Corporation, Marion, Iowa. His responsibilities include the formulation, process development, and scale-up of solid dosage forms. His research interests include process control, fluid bed granulation, and controlled release dosage forms. He received his B.S. in biology from Iowa State University. He is a member of the American Association of Pharmaceutical Scientists and the Controlled Release Society.

JAMES J. SCHIRMER

Mr. Schirmer is currently a Senior Scientist, Process Development, at Schering-Plough HealthCare Products. He has 20 years of experience in formulation, scale-up, and technical support of solid dosage products. His experience includes fluid-bed processing, high shear granulating, film coating, and sugar coating.

ATUL J. SHUKLA

Dr. Shukla is Associate Professor of Pharmaceutics in the Department of Pharmaceutical Sciences, College of Pharmacy, The University of Tennessee, where he has been a member of the faculty since 1989. He was Assistant Professor of Pharmaceutics at the School of Pharmacy, Duquesne University for 3 years after receiving a Ph.D. degree from the University of Georgia in 1985. His research interests include the development of controlled release drug delivery systems, evaluation of excipients in various drug delivery systems, and the use of near infrared spectroscopy for the analysis of pharmaceutical excipients and products. He has published 18 papers in peer-reviewed technical journals.

GREGORY RAYMOND SMITH

Mr. Smith is currently the Vice President of Engineering at Vector Corporation. Previous to this position, Mr. Smith was Vector's Lead Controls Engineer and was responsible for the development and implementation of several control systems throughout the pharmaceutical industry. Mr. Smith received his Bachelor of Science in Electrical Engineering from Iowa State University and is currently pursuing his MBA through the University of Iowa.

ROBERT E. SPARKS

Dr. Sparks was the Lopata Professor of Chemical Engineering at Washington University (St. Louis, Missouri) until 1995. At that time, his microencapsulation laboratory became Particle and Coating Technologies, Inc. In the new company, he is Director, Research and Development, with interests in atomization, particle formation, particle coating, controlled release, and fluid mechanics.

JIM SPENCER

Mr. Spencer is currently the Lead Controls Engineer at Vector Corporation. The Controls Department, which Mr. Spencer manages and leads, is responsible for the development and implementation of all the control systems supplied by Vector Corporation. Mr. Spencer received his Bachelor of Science in Electrical Engineering from Iowa State University and is currently pursuing his MBA through the University of Iowa.

DAVID E. WIGGINS

Mr. Wiggins has over 15 years' experience in the pharmaceutical industry, specializing in analytical method validation and stability of finished dosage forms. He has published several articles on these topics, written monograph procedures for the U.S. Pharmacopoeia, and lectured throughout the world. A member of AAPS and AOAC, he received his B.S. degree from Rhodes College in Memphis, his M.S. from the University of Tennessee in Memphis, and his M.B.A. from the University of Memphis.

1

INTRODUCTION

Atul J. Shukla

The University of Tennessee

Rong-Kun Chang

Shire Laboratories, Inc.

Kenneth E. Avis

The University of Tennessee

THE EVOLUTION OF COATING SOLID DOSAGE FORMS OF DRUGS

Coating is one of the oldest pharmaceutical processes that is still practiced today. Some of the original reasons for coating solid dosage forms, such as pills, granules, and tablets, were to mask the unpleasant taste or smell of a drug or biologically active extract present in the pill or tablet, to prevent corrosive action of the active ingredients on the mucous membranes of the gastrointestinal tract, to facilitate swallowing, to improve the stability of the active component in the solid dosage form, and to improve the aesthetic appearance of the pills or tablets. While the original reasons for applying coatings to these solid dosage forms of drugs have not changed much during the many centuries that the art (or science) of coatings

1

has been practiced, the ingredients used as coatings and the technology for applying them has undergone a tremendous evolution.

Pill coating using mucilage from psyllium seeds was described by Rhazes as early as the 9th century A.D. Besides mucilage, various other types of materials have been used in the past to coat oral solid dosage forms. These include gold leaf and silver leaf, gelatin, salol, tolu, stearic acid, chocolate, shellac, casein, talcum powder, and sugar. Waxes were also often used to coat pills made from poisonous ingredients. This practice was prevalent because the waxes were not soluble in fluids present in any part of the gastrointestinal tract, thus preventing accidental poisoning because the ingredient would not be released. If, however, the poisonous active ingredients were wanted, the pill would be crushed prior to swallowing, in order to release the poison in the body (Cook and Martin 1951; Martin and Cook 1961).

The early coating techniques were extemporaneous and often performed on individual pills by picking them up one at a time, either on the point of a needle or with a pair of forceps, and dipping them in a coating solution. Occasionally, the pills were held at the end of small suction tubes, using vacuum, and one side of the pill was dipped in a coating solution. The process was reversed in order to coat the other side.

The following is a description of several extemporaneous coating techniques employed in the past (Cook and Martin 1951):

- *Gold, silver, and aluminum coating:* A thick mucilage of acacia gum was prepared in a mortar. The pills or tablets to be coated were gently placed in the mucilage and rolled so that the outer surfaces of the pills or tablets were uniformly coated with the mucilage. They were then placed in a globular box (with a smooth inside surface) made from either horn or wood. The box would contain a thin layer of gold or silver leaf. The box was closed, and the pills or tablets were then rotated to facilitate the uptake of the leaf coating on the surface. While gold and silver coating was practiced as early as the 11th century, the process of coating pills with aluminum was performed more recently and primarily replaced silver coating.

- *Pearl coating:* The process of coating solid dosage forms with finely powdered talcum was very popular at one time, and it was often referred to as "pearl coating." In this process, the pills were rolled in acacia mucilage until the surface was uniformly coated. They were then transferred into a smooth globular box containing finely powdered talcum and rolled until a firm white coating was produced. After drying, the talcum-coated pills were polished to obtain a shiny "pearl-like" appearance by rolling them in another box containing paraffin.

Tolu coating: The pills were placed in a porcelain dish containing a thin layer of a solution of equal volumes of ether and tolu tincture and rotated until the surfaces of the pills were uniformly coated with the mixture. The pills were then transferred to another container of similar size and shape and rolled again to remove any excess coating. The pills were then transferred by hand to a third dish whose inside surface was uniformly coated with a thin film of oil. The tablets were rolled in this dish until dry. More than one coat of tolu or oil was often applied to ensure uniform coating.

Gelatin coating: The process of coating pills and tablets with gelatin is also rather old. The method involved sticking a pill on a needle and dipping it in a solution of gelatin. The pills were then dried, and the needle was removed from the pill.

Another procedure for coating pills or tablets with gelatin was to cast a sheet of gelatin on drums. Once the sheets were set and dried, they were then cut to predetermined lengths. Single-cut sheets were placed over each of the two halves of the die with recesses that were similar in shape and size to the pills or tablets. The solids to be coated were placed on top of the gelatin sheet in the bottom recess. The upper half of the die with a gelatin sheet and recess was then placed over the lower half. Sufficient pressure was then applied until the sheets from the two halves were cemented together around the pills or tablets. The coated solids were then cut from the dies, and the excess gelatin was trimmed.

Stearic acid coating: Stearic acid was used as an enteric coating material for pills and tablets. The solid dosage forms were placed and rotated in a dish containing a few drops of saturated ethereal solution of stearic acid. Several layers of coating were occasionally necessary to accomplish uniform coating.

Chocolate coating: The pills or tablets were rolled on a paper saturated with acacia mucilage until the surfaces of the pills were covered with a uniform layer of the mucilage. Powdered cocoa was shaken on the tacky pills, and the pills were then transferred to a warmed porcelain dish and rotated until a smooth coating was produced.

* *Sugar coating:* The pills were rolled on a filter paper saturated with acacia mucilage until the surface was uniformly coated with the acacia and was tacky. The tacky pills were then transferred to a porcelain dish containing mixtures of eight parts of powdered sucrose or lactose and two parts acacia powder. The pills were then rotated in the powder until a firm uniform coating resulted.

It is evident from the aforementioned descriptions that the processes for coating solids were extemporaneous and rather crude. Moreover, only a few solid dosage forms could be coated in one batch, and often the procedures did not result in a uniform, reproducible coating from one unit or batch to another.

The first compressed tablets appeared in the mid-1800s when scientists in England and the United States were awarded patents for inventing the tablet presses (Brockedon 1843; McFerran 1874; Dunton 1876). Soon after the development of the tablet press for the manufacture of compressed tablets, sugar coating of the tablets in coating pans was initiated. This technology was borrowed from the candy industry where it had made significant advancement. The process was introduced in the United States around 1842 (Wiegand 1902) from France. At this point, the extemporaneous sugar coating techniques described earlier became obsolete and were quickly replaced with mechanically revolving coating pans.

In the 1950s, the rotating pans were developed such that their movement could be automatically controlled, thus giving the pans the so-called "jogging cycle." In this cycle, the pans could be started and stopped at any given time by the operators, thus providing a means for obtaining more uniform coating and drying cycles. The coating pans were generally constructed from either copper or galvanized iron. The diameters of these pans ranged from 8 to 60 in. and were rotated on their horizontal axis by a motor. Coating solutions were applied a little at a time to the tablets by ladling (pouring the coating solution from a cup-shaped vessel with a long handle called a ladle) or spraying the material on the rolling tablets in the pan. Drying was performed by heated air that was directed from the front of the pan onto the tablet bed. the air was also exhausted through the ducts positioned in the front of the pans.

Sugar coating using the pan-coating technique allowed formulators to produce shiny tablets of varying colors, since coloring agents could be added to the sugar-coating formulations. Once the coating was dried, the tablets could be polished using waxes to give the tablets a bright, shiny, and elegant appearance.

Although the basic sugar-coating formulations have changed little, the process of sugar coating has undergone a significant change over the years. The traditional copper and galvanized coating pans have been replaced with stainless steel pans containing baffles. The process of applying coating solutions on the tablet surfaces has also been significantly improved by using computer-controlled atomizing systems. The drying process has become more efficient by using completely enclosed pans fitted with more efficient air handling systems.

The coating pans that were used for sugar coating were also used for film-coating tablets with both natural and synthetic film-coating materials. The process of film-coating tablets in pans was much faster than the sugar-coating process.

While significant advances were going on in the traditional sugar- and film-coating technologies, two new technologies—the air suspension technique and microencapsulation—were also being developed. In 1953, Dr. Dale Wurster (1953) patented the method of film-coating tablets and granules with the air suspension technique. In this process, tablets or granules were suspended in the chamber by a strong current of an upward moving airstream. The coating solution was sprayed on the suspended tablets or granules in the chamber. This technique was very efficient and rapid. Moreover, newer coating materials, such as synthetic film-forming polymers, could now be used to coat solids quickly. Further, the method could be adapted to polymers dissolved in volatile solvents, such as acetone, alcohol, and chloroform.

Like the conventional pan-coating technique, the air suspension technique has also undergone significant improvement since it was first introduced. Moreover, newer coating materials, including aqueous dispersions of polymers of ethylcellulose (Aquacoat ECD®[1] and Surelease®[2]), cellulose acetate phthalate (Aquateric CD-910®[1] and Sureteric®[2]), and acrylic acid–based polymers (Eudragit®[3], Eastacryl 30D®[4], and Kollicoat MAE®[5]) have also been introduced recently. Increasingly, these aqueous-based polymeric coatings are replacing the volatile solvent-based coatings because of the potential dangers

---

1. FMC Corporation, Philadelphia, Pennsylvania
2. Colorcon, West Point, Pennsylvania
3. Hüls America, Somerset, New Jersey
4. Eastman Chemical Company, Kingsport, Tennessee
5. BASF Corporation, Mount Olive, New Jersey

of the use of volatile solvents, to both the personnel involved in coating and the environment.

In 1956, the microencapsulation by coacervation process using water-soluble colloids, such as gelatin and gelatin-acacia (gum arabic) mixtures, was developed commercially at the National Cash Register (NCR) Company by Green and Schleicher (1956, 1957). The objective of their research was to encapsulate a dye and develop colorless pressure-sensitive copying paper or carbonless copy paper. The microencapsulation process was later used to coat drug particles in a variety of polymers. Since 1956, many coating materials and microencapsulation processes have been developed by pharmaceutical manufacturers to encapsulate biologically active ingredients. In fact, over the last 45 years, considerable effort has gone into the development of various techniques and processes for microencapsulation, such as solvent evaporation, coacervation by phase separation, and interfacial polymerization. Publications on these processes abound in the literature.

With the rapid advances in both computer hardware and software, the technology of coating is rapidly changing from a totally manual operation of the past to the semiautomatic operation of the present. With this kind of rapid change, the days of totally automated and computer-controlled processes for coating tablets, granules, and particles, either in a coating pan or by the air suspension technology, can be anticipated in the near future. The chapters on sugar coating, film coating, and microencapsulation in this book address some of the issues highlighted above in much greater detail.

CHAPTER CONTENTS

Six chapters follow this introduction. The first three cover the technical principles and their application to the three processes currently used to coat solid dosage forms of drugs. The second three chapters cover other topics closely associated with the coating processes.

Chapter 2, "Sugar Coating," has been written by Eddie Brunson and Jim Schirmer. The authors provide extensive information on the process for the sugar coating of solids, with specific details on the steps involved. Formulations for the solutions or suspensions applied in the various steps of the coating process are given, with descriptions of how these formulations are applied. Problems that may be encountered are discussed, and ways to avoid or overcome them are presented. The processes described progress from manual to highly automated operations, and the equipment required is described in detail. Several photos show aspects of the equipment being discussed. Numerous tables provide information, from a listing of raw materials used and their suppliers to the detailed steps of an example in spray coating, showing the spray volumes relative to times and the intermediate rolling/drying times. It is truly a "how to do it" text. Fifteen references conclude the chapter.

Robert Campbell and Gary Sackett have written a very extensive chapter on the relatively new technology of film coating. To quote from the beginning of their chapter, " . . . aqueous film coating has truly become a science with the recent development of sophisticated coating polymers and automated coating systems." From this perspective, they develop detailed information about coating materials, how to apply them and the equipment required to do so. Twenty-seven figures illustrate the equipment and techniques utilized, including coating pan systems and air suspension systems. Problems encountered in film coating are discussed, and approaches to solving or preventing the problems are presented. The methods required for the automatic application of film coatings, utilizing current technology and computerized control are discussed in detail. It is a very practical presentation with voluminous details on how to be successful in the film coating of solid dosage forms of drugs.

Chapter 4, "Microencapsulation"—the method for coating solid and liquid core particles—has been written by Robert Sparks, Irwin Jacobs, and Norbert Mason. Following an introductory section in which the authors discuss important characteristics of the core particles, they provide a detailed discussion of the nature of coating materials and their formulation for application. An extensive discussion of the equipment required and the procedures utilized is then presented, including spray drying, fluidized bed coating, spinning disk coating, coacervation, interfacial polymerization, and solvent evaporation. Information is also given on the issues involved with scale-up. The final section is a detailed discussion of problems related to microencapsulation. Several figures, which are valuable aids in understanding the processes, present schematics of equipment and the processes being employed. This highly practical and informative chapter concludes with 66 references.

While Chapters 2 to 4 include information on the automation of coating processes, the chapter written by Greg Smith and Jim Spencer (Chapter 5) focuses on the principles of automation and their application to coating systems. The authors discuss the development of automation from pneumatic control to electro-mechanical and digital control using programmable logic controllers, distributed control systems, and personal computers to supervisory control and data acquisition systems. They then discuss in considerable detail specific characteristics of automated coating systems. An interesting section follows in which they identify the distinct phases software products undergo before a coating system is finally completed. A critical aspect follows in which the authors present guidelines and discuss their importance in developing a software system that is "user-friendly" (i.e., users should be able to understand and correctly interpret the information presented on the display screen). The final section discusses how the application of the SP88 standard can be used as a tool for allowing all aspects of a system to be defined in a common form so that all parties involved with a particular process will understand the system required.

The cleaning of equipment used in pharmaceutical processing has become a focus of attention in recent years, both because the industry has come to realize that adequate cleaning of multiple-use equipment is a critical process that should be controlled and validated and because the U.S. Food and Drug Administration has emphasized this procedure in recent inspections. Therefore, Chapter 6 written by William Hall, deals with the specific aspects of cleaning equipment used in the coating of drug dosage forms. After introducing the topic and describing issues in cleaning, he defines a cleaning program as consisting of detailed cleaning procedures, a personnel training program, visual examination of equipment, validation of cleaning procedures, a detailed documentation system, and change control. Each of these components of a cleaning program are discussed in detail. A large portion of the chapter appropriately deals with the evaluation of cleaning, including sampling techniques, analytical methods of testing for residues, and setting limits and acceptance criteria. In the final section of this valuable chapter, the author presents the elements required for a protocol to demonstrate the capability of the cleaning process.

In the final chapter of the book, John Addison and David Wiggins present familiar issues concerning stability and quality control, but as related to coated products. As anticipated, the authors state that stability considerations should include packaging, storage conditions, expiration dating projections, and various parameters, including appearance, friability, dissolution, moisture uptake, and the active ingredient content. Further, with respect to quality control, the authors point out that planning for the quality of the product begins in the development stage. The process selected is chosen to be the best to achieve the desired endpoint. It is verified by quality control procedures that show it does accomplish the intended design. the various aspects of a quality control program are discussed, with specific applications to coated pharmaceutical products. One of the practical inclusions is a sample "Process Quality Control Defect and Audit Report" form with a list of potential critical, major, minor, and branding defects. As with each of the other chapters, this one is a practical, applied treatment of the topic relevant to coated pharmaceutical dosage forms.

The topics presented in this book have been written by experts in their respective fields and have been designed to provide practical guidelines for personnel responsible for the manufacture of coated dosage forms of pharmaceutical products. It is with high expectations that this book will prove to be very useful to such professionals that we publish this third volume in the *Drug Manufacturing Technology Series*. Other volumes will soon follow to enhance the understanding of drug manufacturing technology in selected areas.

REFERENCES

Brockedon, W. 1843. British Patent 9977.

Cook, E. F., and Martin, E. W. 1951. *Remington's Practice of Pharmacy,* 10th ed. Easton, Penn., USA: Mack Publishing Company, pp. 1394–1400.

Dunton, J. 1876. U.S. Patent 174,790.

Green, B. K., and Schleicher, L. 1956a. U.S. Patent 2,730,456.

Green, B. K., and Schleicher, L. 1956b. U.S. Patent 2,730,457.

Green, B. K., and Schleicher, L. 1957. U.S. Patent 2,800,457.

Martin, E. W., and Cook, E. F. 1961. *Remington's Practice of Pharmacy,* 12th ed. Easton, Penn., USA: Mack publishing Company, pp. 476–494.

McFerran, J. A. 1874. U.S. Patent 152,666.

Wiegand, T. S. 1902. *Am. J. Pharm.* 74:33.

Wurster, D. E. 1953. U.S. Patent 2,648,609.

2

SUGAR COATING

Eddie L. Brunson
James J. Schirmer

Schering-Plough HealthCare Products

HISTORY

Coatings are not specific to today. References to drug coating date back 10 centuries (Lehman and Brögmann 1996; Porter 1981; Seitz et. al. 1986; Sonndecker and Griffenhagen 1990). The French were instrumental in developing sugar coatings for pharmaceuticals in the mid-1800s (Lehman and Brögmann 1996; Seitz et. al. 1986). The 19th century also saw the advent of pill coating in rotating pans (Robinson 1980; Seitz et. al. 1986).

For the most part, pan coating processes were guarded secrets and required skilled workers to obtain elegant, sugar-coated tablets. The primary materials utilized were common and readily available items, such as sugar, flour, acacia, and gelatin. These materials are still used today with other adjuvants. Sugar coatings normally doubled the initial weight of the core tablet and were lengthy processes, sometimes requiring several days.

Pan film-coated tablets were introduced in 1953. Fluid bed film-coated tablets followed shortly thereafter (Lehman and Brögmann 1996; Seitz et. al. 1986). These processes employed various types of polymers as the coating material, required shorter process times, and removed most of the art required by sugar coating. However, these improvements were offset somewhat by the higher cost of the polymer used. Although shorter process times and better process controls were made possible (Wood and Scarpone 1983), properly sugar-coated tablets still have a more aesthetic and pleasing appearance. Many products are still sugar coated today.

In the 1950s, the pear-shaped and oval coating pan designs of the 1930s and 1940s were replaced by side-vented (perforated) pans (Lehman and Brögmann 1996; Seitz et.al. 1986). Improvements continue in the area of sugar and film coating processes to eliminate the guesswork of early coating artisans.

APPLICATIONS

Although coatings add an additional manufacturing step to the production process, they may provide one or more of the following product attributes:

- Mask the unpleasant taste of the drug(s).
- Allow the tablet to be swallowed more easily.
- Protect the active drug from the environment.
- Control the drug release site.
- Restrict drug-excipient interactions.
- Make the product aesthetically pleasing.
- Provide a means of identification.

The rationale for deciding to coat a tablet needs to be weighed against the added labor cost. Masking bitter tasting drugs, the instability of the active constituent, or the desired site of action may leave no other alternative than coating.

The following discussions will focus on the conventional sugar-coating process. An adaptation of an automated sugar-coating process will be given as an example.

CONVENTIONAL SUGAR COATING

Sugar coating is used in immediate release applications to mask drug taste or to provide an aesthetic appeal. It should be understood that sugar coating will add some time to the overall disintegration of the tablet and may impact drug dissolution (Khalil et. al. 1991). This effect should be taken into account when formulating the core and in sugar coating to meet U.S. Pharmacopeia (USP) disintegration and dissolution requirements for compendial products.

For an enteric or sustained-release sugar-coated product, the formulation problem may become more complex to meet USP tablet disintegration and dissolution specifications. The selection of the core tablet and coating materials becomes more important for these applications and requires proper evaluation to assure long-term chemical and physical stability.

Sugar coatings are essentially aqueous based, unless a seal coat (e.g., alcoholic confectioners glaze) is required to protect the core tablet from water used in the sugar-coating process. Although the sugar coating process is lengthy and labor intensive, coating materials are inexpensive and readily available.

Process Overview

The conventional sugar-coating process is a manual operation that includes the following steps:

- Seal coating (optional).
- Subcoating (optional).
- Grossing coats.
- Smoothing coats.
- Color coats.
- Polishing.
- Imprinting (optional).

Simply stated, each step consists of one or more applications of aqueous solutions poured onto the surface of a preheated tablet bed as it tumbles in a rotating pear-shaped or cylindrical pan, sometimes fitted with baffles to aid uniform blending of the tablets. For each application, the tablets are wetted with a measured volume of the specified coating solution, allowed to tumble to distribute the solution evenly over the tablet bed, and then dried with conditioned air. This basic process is repeated throughout the different phases of coating.

The primary purpose of seal coating is to protect the core tablet if it contains a water sensitive drug and to prevent the tablet from absorbing water, softening, and initiating disintegration of the core during the sugar-coating process. The subcoating process binds powders to the core and significantly increases the weight of the tablet, thus allowing a fast buildup of the sugar coating to cover and round off the edges of the core tablet. Grossing coats are suspensions of subcoating powders that can be used for the same purpose as the subcoating. Smoothing coats are applied to complete the filling in of surface imperfections and to prepare the tablet for color coating. The color-coated product is then polished to a glossy, elegant finish. Polishing yields the desired high gloss aesthetic appearance characteristic of sugar-coated products. The polished tablet can be imprinted with a distinctive ink logo for product identification purposes.

Core Tablet Properties

Core tablets for coating can be of many shapes and sizes and possess different physical characteristics. Tablet shapes can range from standard round to oval to capsule with thick or narrow sidebands.

Selecting the proper tooling for tablet compression can minimize subsequent problems during the sugar-coating process. Wide, flat tablet surfaces should be avoided, because they may promote twinning (i.e., the sticking together of tablets in the coating process). For example, the twinning of tablets and straight-sided capsule shapes can be prevented by using tablet tooling that yields curved tablet surfaces and capsule-shaped tooling designed to eliminate straight sides. The width of the tablet sideband is an important parameter to be considered. For example, if this band is too wide, covering and rounding this side of the tablet with either subcoating or grossing coats will be a problem. A thin sideband minimizes these issues.

Physically, the tablets must be able to withstand the abrasive forces of the coating process. A soft tablet with high friability will erode and crumble when exposed to the rigorous tumbling action in a coating pan. In general, tablets with a hardness of 6 kp or greater and a friability of less than 0.5 percent should coat with few, if any, problems. Ideally, formulators should maximize tablet hardness and minimize tablet friability, while, at the same time, maximizing surface curvature with minimal sideband width to achieve the best coating shape.

The inclusion of various functional tableting excipients—compression aids, fillers, disintegrants, and lubricants—may adversely affect the adherence of the coating and the ability of the core tablet to withstand heating during processing. These parameters should be evaluated to establish the ruggedness of the core formula.

Equipment

The equipment used for sugar coating consists of a rotating drum capable of producing a continuously tumbling tablet bed. Such equipment can range from very elaborate, fully automated coating systems to very simple, pear-shaped pans. Ancillary equipment that is necessary for this process includes adequate dry air and exhaust sources and a means of applying solutions and/or powders to the tumbling tablet surface. There are two basic types of panning equipment: conventional pans and side-vented pans.

Conventional Pans

Conventional pans are coating pans in which drying air flows onto the surface of the tumbling tablet bed. Moist, dust-laden air is removed via an exhaust duct. Conventional pans either rotate about a horizontal axis or at an approximate 25° angle (Sonndecker and Griffenhagen 1990). Both types can control the speed of rotation. The oldest pans found in pharmaceutical production were pear-shaped copper pans that rotated at an approximately 25° angle. Stainless steel pans modeled after these pans can still be found in many pharmaceutical operations (e.g., Stokes® coating pans). A variation of this type of pan is the Lakso AR® angular welded stainless steel pan designed to improve the mixing of the tablet bed during tumbling in the pan.

Air supplies were usually simple ducts capable of blowing heated air onto the tablet surface. An exhaust duct would then remove the humid, dust-laden air.

The major disadvantage of angular rotation coating pans is poor mixing action of the tablet bed. As a result, coating solutions and powder applications may not be uniformly dispersed throughout the tablet bed, thus leading to poor coating uniformity.

The next generation of conventional coating pans was designed to improve the mixing and tumbling action during coating. These were angular coating pans rotating on a horizontal axis[1] (see Figure 2.1). These pan designs also incorporated mixing baffles to further improve mixing and included improvements in the ducting of drying air and exhaust to improve drying uniformity and efficiency.

The other type of coating pans, perforated pans, were designed to deliver improved drying efficiencies necessary for the application

---

1. Pellegrini®, Nicomac,Inc. (USA), Englewood, New Jersey; Driacoater®, Driam USA, Inc., Spartanburg, South Carolina.

Figure 2.1. Pellegrini® conventional coating pan. Courtesy of Schering-Plough HealthCare Products.

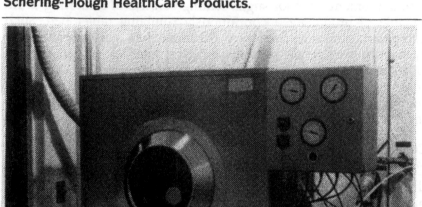

of aqueous film coatings, but have been used successfully for sugar coating[2] (see Figures 2.2, 2.3, 2.4, and 2.5).

Perforated Pans

Side-vented pans are similar in shape to the horizontal axis conventional pans. The primary difference is in the manner in which air is exhausted from the pan. Perforated pans have a perforated surface through which the air is exhausted from the pan directly through the tumbling tablet bed.

Dosing Equipment

Dosing equipment in its simplest form is a manual operation, (i.e., solutions and powders are applied manually to the tablets using ladles, volumetric cups, and scoops). Automation of solution addition by pumping and the addition of powders with powder feeders can be accomplished by using appropriate process controllers.

2. Accela-Cota®, Thomas Engineering Inc., Hoffman Estates, Illinois; Hi-Coater®, Vector Corporation, Marion, Iowa; GCX Pan Coater®, Glatt Air Techniques Inc., Ramsey, New Jersey.

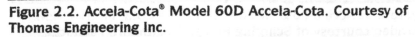

Figure 2.2. Accela-Cota® Model 60D Accela-Cota. Courtesy of Thomas Engineering Inc.

Air Handling

Successful sugar coating requires a reliable supply of conditioned air with an adequate means of temperature control. An experienced operator can manually apply solutions and powders to the tablet bed and can react to changing environmental conditions to achieve an elegant sugar coating. However, an automated system requires a more exacting control of air supply and solutions. Air supply humidity, temperature, and volume need to be controlled within tight tolerances to achieve reproducible and elegant sugar coatings. Maintaining controlled volumes of exhaust air is as critical to the drying dynamics in the coating pan as is the air supply; it needs to be adequately controlled to ensure reproducible sugar coating.

Figure 2.3. Accela-Cota® Model 24 Accela-Cota, interior view. Provided courtesy of Schering-Plough HealthCare Products.

Conventional Sugar Coating Steps

The following conventional sugar-coating process illustration uses a Pellegrini® MC 25 (24 in.) coating pan. The tablet load and process parameters are listed in Table 2.1. Materials (grades and vendors) discussed in the text are listed in Table 2.2.

Seal Coating

Seal coating is an optional step but is usually required if the drug is water sensitive or if the tablet core is susceptible to water absorption

Figure 2.4. Hi-Coater® coating pan. Courtesy of Schering-Plough HealthCare Products.

Table 2.1. Conventional Sugar-Coating Pan Equipment Parameters

Pan Type	24 in. Pellegrini® with baffles
Airflow	Inlet: 200 CFM Exhaust: 300 CFM
Pan Speed	10–15 rpm
Inlet Air Temperature	40–60°C
Pan Load	10 kg
Tablet Weight, Size, Shape	250 mg; 3/8 in. standard round, convex

from the coating process. Water absorption results in undesirable softening and the initiation of core disintegration.

Materials providing a water repellent coating include alcoholic solutions of shellac, zein, or polyvinyl acetate phthalate (PVAP). Waxed and dewaxed shellac have been used for sealing purposes.

Figure 2.5. Signature Series GCX® pan coater. Courtesy of Glatt Air Techniques.

However, it should be noted that after extended storage, shellac polymerization may result in an undesirable increase in tablet disintegration time (Porter 1981). Waxed shellac reportedly does not undergo appreciable polymerization on extended storage (Mantrose et al.). Polymerization has not been reported with zein (Porter 1981; Seitz et. al. 1986). PVAP, an enteric coating polymer, should be used sparingly, since it can also impact disintegration of the product.

Table 2.2. Sugar Coating Raw Materials

Material	Grade	Suppliers
Acacia	Spray Dried NF	Colloides Naturels Gumix Intl Inc. Penta Mfg. Co. TIC Gums Inc.
Calcium Carbonate	USP Precipitated, Light	Rhone-Poulenc Specialty Minerals Inc.
Calcium Sulfate Dihydrate	NF	EM Industries Whittaker Clark & Daniels Inc.
Carnauba Wax	NF	Centerchem Inc. Penta Mfg Co. Strahl & Pitsch Inc.
Cellulose Acetate Phthalate	NF	Eastman Chemical Company FMC Corp./Pharmaceutical Division
Coloring Agents	FD&C or D&C	Colorcon Crompton & Knowles Warner Jenkinson
Isopropyl Alcohol	USP 99%	Ashland Chemical
Kaolin	USP	Penta Mfg Co. Spectrum Quality Products Inc. Whittaker Clark & Daniels Inc.

Continued on next page.

Continued from previous page.

Material	Grade	Suppliers
Polyvinyl Acetate Phthalate (PVAP)	NF	Colorcon
Shellac	NF, Pharmaceutical Glaze	Colorcon Mantrose Bradshaw—Zinsser Group
Starch	NF	American Maize Products Grain Processing Corp National Starch Roquette America
Sucrose	NF	Ashland Chemicals Corp./Fine Ingredients Division Paular Corp.
Talc	USP	Ashland Chemicals Corp./Fine Ingredients Division Luzenac America Whittaker Clark & Daniels Inc.
Titanium Dioxide	USP	Ashland Chemicals Corp./Fine Ingredients Division Crompton & Knowles Warner Jenkinson Whittaker Clark & Daniels Inc.
White Wax	NF	Amoco Petroleum Products Penta Mfg Co. Strahl & Pitsch Inc.
Zein	NF	Atomergic Chemetals Corp.

The sealing process normally incorporates one or two applications of the seal coat solution. A dusting powder (e.g., talc) is added to prevent sticking of the coated cores after each seal coat application. Typical seal coat formulations and process parameters are given in Table 2.3.

In a manual seal-coating operation, the coating solution is poured evenly and slowly over the tumbling tablet bed (preheated to 40°C). It is allowed to distribute throughout the tablet bed with exhaust air on. The pan speed is set at 10–15 rpm. As the alcoholic solvent dissipates, the coating may go through a tacky phase, where tablets will begin sticking together and adhering to the interior of

Table 2.3. Conventional Sugar Coating: Sealing Solutions

Materials	Formulations (as %w/w concentration)		
	1	**2ª**	**3**
Solutions			
1. Shellac	40	—	—
2. Polyvinyl Acetate Phthalate	—	30	—
3. Zein	—	—	20
4. Alcohol, 95%	60	70	64
5. Purified Water, USP	—	—	16
Solids, %w/w	40	30	20
Solution Application	10–15 mL per kg of cores		
Solution Temperature	60–65°C		
Inlet Air Temperature	60°C		
Number of Applications	2 or more		
Dusting Powder			
1. Talc, USP	Apply 10–20 g per kg of cores		
Tablet Weight Gain	~1–2%		

a. Opaseal®, Colorcon, Inc., West Point, PA.

the coating pan. If this occurs, dusting powder is added to prevent the tablets from sticking. After the dusting powder has been evenly distributed, dry air at 60°C is applied for 10–15 min to dry the seal coat. Additional applications of the seal coating, if needed, are applied in the same manner.

The main problems encountered during seal coating are tablets sticking together and to the pan, and too many seal coat applications that may lead to increased disintegration time. Sufficient dusting powder should be applied to minimize tablet sticking. Seal-coated cores should be subjected to disintegration testing at different application levels to establish effective sealing with minimal moisture uptake and prolongation of the disintegration time.

Subcoating

The purpose of subcoating is to round off sharp edges of the tablet and provide a good substrate for the subsequent smoothing coats. This step is optional. Tablet weight may increase by 50 percent during this phase. Binder solution materials include sugar, water, gelatin, or acacia. Dusting powder blends may utilize calcium carbonate, calcium sulfate, talc, titanium dioxide, and acacia. Typical subcoating binder solutions, process parameters, and dusting powder blend formulas are given in Table 2.4.

The binder solution is prepared by adding water to a suitable container. The binder is added to the water and dispersed with heating until dissolved (Burroughs 1996). Sugar is added and dissolved. The resultant solution is cooled to approximately 60°C for the coating process. If necessary, additional water may be added to adjust the solution to a lower viscosity, which will allow the binder solution to spread evenly and uniformly throughout the tablet bed.

A typical subcoat procedure is as follows. The binder solution is applied slowly over the tumbling tablet bed. The exhaust and inlet air supply are off, and the pan rotation is set at 10–15 rpm. When the solution is evenly distributed over the tablet bed, the dusting powder is added and allowed to spread and adhere to the tablets. After the powder has adhered to the tablets, drying is initiated with the exhaust and inlet air on. The drying air temperature is set at approximately 60°C. The drying cycle is usually 2–5 min. Subsequent applications are made in the same manner until the rough edges of the core tablet are rounded.

This process can be messy due to the application of a tacky binder solution and large powder volumes. The major problems associated with subcoating are as follows:

**Table 2.4. Conventional Sugar Coating: Subcoating
Solutions/Dusting Powders**

Materials	Formulations (as %w/w concentration)			
	1	2	3	4[a]
Solutions				
1. Sugar	61.8	50	50	45
2. Gelatin	5	—	4	6
3. Acacia	—	8	4	8
4. Purified Water, USP	33.2	42	42	41
Solids, %w/w	66.8	48	58	59
Solution Application	10–30 mL/kg of cores			
Solution Temperature	60°C			
Inlet Air Temperature	60°C			
Number of Applications	2–10			
Dusting Powders				
1. Calcium Sulfate	100	47.6	—	—
2. Calcium Carbonate	—	—	95	—
3. Talc	—	47.6	—	61
4. Titanium Dioxide	—	4.8	—	1
5. Acacia, powdered	—	—	5	—
Powder application amount	10–50 g/kg of cores			
Tablet Weight Gain	~25–50%			

a. Porter (1981)

- Twinning (sticking of two or more tablets).

- Production of rough tablet surfaces.

- Variability in weight and thickness uniformity of the powder-coated cores.

- Rough powder buildup on the surface of the coating pan.

Twinning may result from localized overwetting of the bed, a high-viscosity binder solution, insufficient application of dusting powder, or a combination of these. Remedies include reducing the solution volume, diluting the solution to reduce its viscosity, increasing the powder application, or a combination of these.

Excessive dusting powder application will result in rough and lumpy tablet surfaces. A decrease in the amount of powder will help to control the lumpy surface texture, but it will not eliminate the problem.

Excess variability in weight and nonuniformity of thickness results when the aforementioned problems exist. Some variation is expected due to the uneven transfer of solution and powder on tablet surfaces during coating. Completely eliminating these problems is not possible. However, careful applications will minimize these problems.

The buildup of a rough powder surface on the coating pan is due to the transfer of binder solution from the tablet bed to the coating pan. Subsequent powder applications adhere not only to the tablets but also to the tacky pan surface. This problem cannot be eliminated but may be reduced by adjusting the solution applications to minimize the transfer of binder solution from the tablets to the pan. Rough powder buildup on the interior pan surface may necessitate cleaning of the pan prior to continuation of the coating process.

Grossing Coats

Grossing coats are suspensions of subcoating powders that may be substituted for the subcoating process. Grossing suspensions can minimize the above-mentioned problems associated with subcoating powders.

The materials used are similar to those listed in the subcoating section. Typical formulas and process parameters for this application are given in Table 2.5.

Solution preparation is similar to that of the binder solution for subcoating. Water is added to a suitable container. The binder is added to the water and dispersed with heating until dissolved. Sugar

Table 2.5. Conventional Sugar Coating: Grossing Solutions

Materials	Formulations (as %w/w concentration)		
	1	2	3[a]
Solutions			
1. Calcium Sulfate	—	32	—
2. Calcium Carbonate	15	—	20
3. Talc	2	—	12
4. Titanium Dioxide	—	—	1
5. Acacia, powdered	—	3	2
6. Sugar	63	45	40
7. Purified Water, USP	20	20	25
Solids, % w/w	80	80	75
Solution Application Amount	10–30 mL/kg cores		
Solution Temperature	60°C		
Inlet Air Temperature	60°C		
Number of Applications	15–45		
Tablet Weight Gain	~20–60%		

a. Porter (1981)

is added and dissolved in the solution. The resultant solution is cooled to approximately 55°C for this coating operation. Additional water may be added to reduce the viscosity of this solution to aid uniform distribution over the tablet bed.

Applications of the grossing suspension are added with the exhaust and inlet air supply off. The solution is applied evenly over the bed at a pan speed of 10–15 rpm. After the suspension is thoroughly distributed, the inlet and exhaust air are turned on and drying is initiated with warm air at 60°C. Drying continues for approximately 2–5 min or until dusting is noted in the pan. The inlet and exhaust air are then shut off and the next application is made. Initial applications will produce a rough textured surface.

When the sharp tablet edges become rounded, the suspension viscosity may be lowered to allow low spots to fill in and high points to dissolve in preparation for smooth coating. The coated tablets will have a rough surface texture at the end of this process.

The major problems encountered during grossing coat applications are the development of an excessive amount of pareils and buildup of dried rough coating on the pan. Pareils are small balls of the suspension solids that form in the pan and are caused by an excessive addition of coating solution. Pareils may attach themselves to tablets with subsequent suspension applications and form a bump on the tablet surface. This problem may be avoided by reducing the volume of solution to an amount that will sufficiently wet the tablet bed with minimal transfer of the suspension solution to the coating pan.

The grossing process may be lengthy and result in the buildup of solution solids on the pan. It may be necessary to remove the coated tablets at the end of this process to clean the pan. Coated tablets should be transferred to open trays at a depth of less than 4 in. in an air-conditioned area (70–75°F/40–50 percent relative humidity) to dry for at least 12–15 h before continuing with the next step—smooth coating.

Smooth Coating

The purpose of smooth coating is to smooth the rough tablet surfaces that have been formed as a result of the subcoating and gross-coating operations. The goal is to provide a smooth base for the color coats to be applied.

The materials generally used for the smoothing operation consist of simple sugar syrup ranging in content from about 60 to 70 percent w/w sugar (see Table 2.6 for typical formulas and process parameters). The percentage of sugar necessary for this coating step depends on the amount of smoothing that is required. If large surface pits are to be filled in, higher percentages of sugar syrup solids are required. However, if a relatively smooth surface is to be improved, a lower percentage of sugar syrup solids is used to dissolve high spots on the tablet surfaces and fill in minor surface irregularities.

The application of smoothing syrups is generally the same as with the grossing coat application. However, the tablets are allowed to tumble wet longer, 1 to 3 min after solution application, before warm air at 60°C is introduced. This longer wet roll time results in a smoother surface due to the partial dissolution of high spots on the tablet surfaces. During the smoothing process, decreasing

Table 2.6. Conventional Sugar Coating: Smoothing Solutions

Materials	Formulations (as %w/w concentration)	
	1	2
Solutions		
1. Sugar	60	70
2. Purified Water, USP	40	30
Solids, % w/w	60	70
Solution Application Amount	10–20 mL/kg of cores	
Solution Temperature	60°C	
Inlet Air Temperature	60°C	
Number of Applications	5–15	
Tablet Weight Gain	~5–10%	

amounts of solution are applied as the surfaces become smoother and easier to wet due to the decreased surface area. Drying rates during smoothing should be slower than during grossing applications in order to allow a fine crystalline surface to develop.

The major problem occurring during smoothing operations is the development of rough tablet surfaces, which could result from the application of an excessive amount of sugar syrup, inadequate wet roll times, or fast drying rates. An optimum amount of sugar syrup is necessary to wet the tablet surfaces thoroughly and evenly. If too much sugar syrup is added, the wet roll time will not be long enough to dissolve the high spots remaining from the grossing applications. A drying rate that is too fast causes the sugar to crystallize into a larger crystalline structure with a rough surface texture.

Color Coating

Color coating is the application of a sugar syrup containing dispersed pigments of lake dyes and titanium dioxide. The goal of color coating is to achieve an evenly colored, smooth surface suitable for polishing to a glossy, elegant finish. The materials used to color coat tablets include sugar, water, lake dyes, titanium dioxide, and modifiers such as gelatin or acacia. Using color dispersions from companies experienced in supplying colors to the pharmaceutical industry precludes having to make pigment dispersions suitable for addition to the sugar syrup.

Solution preparation is similar to that for the previous coating steps. Water is added to a suitable container. Acacia and/or gelatin (if desired) is added to the water and dispersed with heating until dissolved. Sugar is added and dissolved. The resultant solution is cooled to approximately 50°C, and the color dispersion is added and mixed until uniform. Typical color coat solutions and process parameters are listed in Table 2.7.

The application of color syrup should follow the same guidelines as for smooth coating–a quantity of syrup should be applied so as to allow the complete wetting of the tablet bed with rolling for 1 to 3 min before applying warm air at 50°C.

Potential problems associated with color-coating processes include uneven color development, chipping, rough coatings, and nondrying coatings. Uneven color is evident by either a mottled appearance of the surface or a failure of the color to build on the edges of the tablets. Mottling is caused by insufficient color syrup, resulting in uneven coverage of the tablet surfaces, or insufficient drying of each color application. Tablets should be allowed to dry until an even, slightly dusty, dry appearance is evident before the next color application is performed. Excessively high quantities of color syrup will result in a "washing" of the tablet edges. This is remedied by lowering the amount of each application of color syrup until the edges are covered. Washing of the edges can also be caused by too much dilution of the color syrup.

Chipping of the tablet edges is the result of excessively high pan speeds, excessive drying, or improper color syrup formulation. The linear speed of the pan surface should not exceed 200 ft/min. Excessive drying can cause the edges of the tablets to wear or chip from attrition. Excessive pigment loading in the color syrup can result in a weak coating that is prone to chipping. The addition of gelatin or acacia modifies the sugar syrup and can strengthen the coating.

Rough coatings are the result of the same causes as in the smoothing stages. Proper dose volumes, adequate roll times, and proper drying are required for a smooth, elegant finish. Decreasing the volume of color syrup toward the end of the coloring process is necessary to finish the tablets properly. The final few applications of color syrup should be of a small enough volume to just wet the surfaces of all the tablets before they are dried with exhaust air, thus achieving a very fine, smooth surface.

Table 2.7. Conventional Sugar Coating: Color Solutions

Materials	Formulations (as %w/w concentration)	
	1	2
Solutions		
1. Sugar	66.6	66.6
2. Gelatin	0.5–2.0	—
3. Acacia	—	0.5–2.0
4. Color Dispersion (70% solids)	10	10
5. Purified Water, USP	21.4–22.9	21.4–22.9
Solids, % w/w	74.1–75.6	74.1–75.6
Solution Application Amount	10–20 mL/kg of cores	
Solution Temperature	50°C	
Inlet Air Temperature	50°C	
Number of Applications	10–20	
Tablet Weight Gain	~5–10%	

The failure of a coating to dry is a result of inversion of the sugar-coating solution. Inversion is due to the acid hydrolysis of sucrose. Inverted sugar syrup will not crystallize and causes a tacky, nondrying coating. Keeping the color coat solution at approximately 50°C during use will prevent inversion. Once a syrup inverts, it is unsalvageable; however, in the authors' experience, inversion is a very rare occurrence. In general, it is prudent to make and use this solution on the same day.

Finish Coating

The purpose of finish coating is to provide an extremely smooth surface suitable for polishing. Additionally, the finish syrup helps to seal in the color, thus preventing the transfer of color to the polishing pans. Finish syrup is a 60 percent w/w sugar syrup. After the final color application, one or two measured volumes of finish syrup are applied to the tablets (see Table 2.8 for a typical formulation and

Table 2.8. Conventional Sugar Coating: Finish Solution

Materials	Formulation (as %w/w concentration)
	1
Solutions	
1. Sugar	60
2. Purified Water, USP	40
Solids	60% w/w
Solution Application Amount	5 mL/kg of cores
Solution Temperature	Room Temperature
Inlet Air Temperature	None Applied
Number of Applications	2 or more
Tablet Weight Gain	~0.5–1.0%

process parameters). As in the last color applications, the quantity of syrup should be just sufficient to wet the tablet surfaces. The tablets should be allowed to dry only until the surfaces lose their shiny wet appearance, but not so long that dusting of the surfaces is evident. After the final finish application loses its shiny wet appearance, the pan speed should be slowed to allow the tablets to dry very slowly, thus forming an extremely smooth surface finish. Dried tablets should be removed from the pans and allowed to dry thoroughly before polishing (e.g., overnight in an air-conditioned room [70–75°F/40–50 percent relative humidity] on trays loaded to a depth of less than 4 in.

Polishing

The last step in achieving an elegant sugar coat is polishing of the tablet surfaces. Polishing is the application of waxes to the tablet surface via an organic solvent carrier and is usually performed in canvas-lined, pear-shaped coating pans. The canvas lining provides the friction necessary to polish the waxed tablet surfaces to a high gloss.

Refined white wax and carnauba wax are used in proportions of about 1:1. The waxes are dissolved in appropriate organic solvents, or finely divided waxes can be slurried in an alcoholic carrier. Typical formulas and process parameters are listed in Table 2.9. An amount of polishing solution to wet the tablet surfaces completely is applied to the tumbling tablet bed in a polishing pan. The pan speed is more rapid than during sugar coating (to allow the tablets to tumble properly)—usually about 30 rpm for a 42 in. pan. The waxed tablets are allowed to tumble until glossy, usually about 20 min.

If too much wax solution is added, an uneven buildup of wax on the tablet surface may result; however, most polishing problems stem from rough surfaces due to improper smooth and color coating. Any roughness on the tablet surface will become more evident

Table 2.9. Conventional Sugar Coating: Polish Solutions

Materials	Formulations (as %w/w concentration)	
	1 [a]	2 [a]
Solutions		
1. Carnauba Wax	6.25	6.25
2. White Wax	6.25	6.25
3. Petroleum Ether	93.75	—
4. Alcohol, SD3A	—	93.75
Solids, % w/w	12.5	12.5
Solution Application Amount	4–8 mL/kg of cores	
Solution Temperature	60°C	
Inlet Air Temperature	Room Temperature	
Number of Applications	1	
Tablet Weight Gain	~0.1–0.2%	

a. Use steam heat only, properly grounded metal containers, and explosion proof equipment.

after polishing. Unfortunately, this type of problem cannot be remedied by a change in the polishing procedure and may require rework.

Rework

The rework of sugar-coated tablets is not recommended. If necessary, it requires removal of the rough tablet surfaces. This can be accomplished by reapplying color coats and finish coats as needed to achieve the desired smooth surface for polishing. The procedure to perform these steps is as previously described.

Imprinting

Imprinting or branding equipment[3] (see Figures 2.6 and 2.7) puts an identifying mark or logo on the surface of the finished, polished, sugar-coated tablet. In the branding operation, a pharmaceutical grade edible ink is applied to a gravure (engraved) roller into which the desired logo is etched. The excess ink is removed from the roller by a scraper blade, and the ink remaining in the etched cavities is transferred to a rubber roller and then onto the surface of the tablet.

The control of tablet thickness is critical to proper branding operations; therefore, most tablets to be branded undergo a sorting operation. Sorting utilizes rotating rollers over which the tablets are distributed. The rollers are spaced to remove under- and oversized tablets. Excessive thickness variation can result in incomplete or smudged brands, which is alleviated by sorting.

Potential problems with branding include off-center branding and incomplete and blurred or smudged brands. Off-center branding is corrected by the proper setup and indexing of the branding equipment. Incomplete branding is either caused by tablets being too thin or ink of an improper consistency (i.e., ink drying before transfer to the tablet surface). Smudging of the brand is usually caused by the ink drying too slowly. Ink consistency is corrected by dilution with the appropriate solvent for the ink formulation in use (Hayes 1996). Pharmaceutical grade branding ink is available from companies that supply certified dyes to the pharmaceutical industry.

Summary

Conventional sugar coating is a labor-intensive process. However, if the operation is properly performed, coatings can achieve aesthetic

---

3. Ackley Ramp and Drum Printers, Ackley Machine Corporation, Moorestown, New Jersey.

Figure 2.6. Tablet ramp printer. Courtesy of Ackley Machine Corporation.

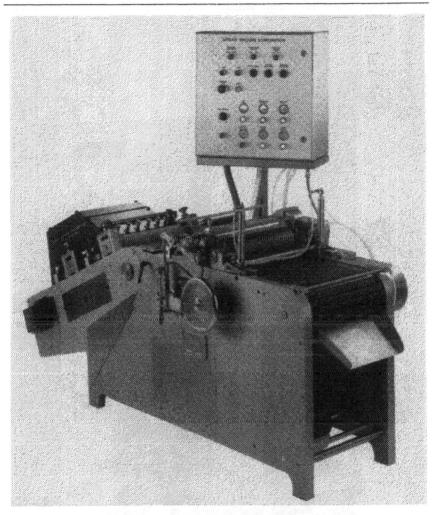

elegance, resulting in an evenly colored, smoothly finished, and highly polished sugar-coated tablet.

Conventional Tablet Sugar-Coating Process Example

The following conventional sugar-coating illustration utilizes a Pellegrini® Model MC-25 (24 in. pan). The tablet load, formulas, and

Figure 2.7. Tablet drum printer. Courtesy of Ackley Machine Corporation.

process parameters are listed in Table 2.10. The coating parameters are given in Table 2.11.

Grossing Coats

This example does not employ a seal coat or subcoating-dusting powder applications to cover the tablet edges. Instead, the grossing coats applied directly to the core tablet to round off its sharp edges

Table 2.10. Manual Sugar Coating Process Example: Equipment/Parameters

Tablet Cores—Conventional Pan	
Tablet:	3/8 in. diameter, convex face core
	Core Tablet Weight: 205 mg
Solutions:	Grossing Coat: Formula 1 in Table 2.5
	Color Coat: Formula 1 in Table 2.7
	Finish Syrup: Formula in Table 2.8

Coating Conditions:

Pellegrini Coating Pan	Model MC-25
Loading Volume	15.0 kg
Pan Rotation Speed	25 rpm
Ambient Relative Humidity	30%
Drying Air Temperature	60–75°C
Drying Air Volume	250 CFM
Exhaust Air Volume	300 CFM
Solution Temperature	ambient to 60°C

Results:

Coating Process Time	216 minutes
Amount of Solution Used	8.500 L Grossing Solution
	2.575 L Color Syrup
	0.090 L Finishing Syrup
Coated Tablet Weight	380 mg (theoretical)
Weight Increase	85% (theoretical)

Table 2.11. Manual Sugar Coating Process Example: Data

Tablet Cores—Conventional Pan					
Solution Type	Coat Number	Solution Volume (mL)	Wet Roll (min)	Dry Time (min)	Drying Air Temp °C
Grossing Syrup					
	1	200	1.0	3.0	65
	2	225	1.0	3.0	65
	3	225	1.0	3.0	65
	4	225	1.0	3.0	65
	5	225	1.0	3.0	65
	6	225	1.0	3.0	65
	7	225	1.0	3.0	65
	8	225	1.0	3.0	65
	9	225	1.0	3.0	65
	10	250	1.0	3.0	65
	11	250	1.0	3.0	65
	12	250	1.0	3.0	65
	13	250	1.0	3.0	65
	14	250	1.0	3.0	65
	15	250	1.0	3.0	65
	16	250	1.0	3.0	65
	17	250	1.0	3.0	65
	18	250	1.0	3.0	65
	19	250	1.0	3.0	65
	20	250	1.0	3.0	65
	21	250	1.0	3.0	65
	22	250	1.0	3.0	65
	23	250	1.0	3.0	65
	24	250	1.0	3.0	65
	25	250	1.0	3.0	65
	26	250	1.0	3.0	65
	27	250	1.0	3.0	65
	28	250	1.0	3.0	65

Continued on next page.

Continued from previous page.

Solution Type	Coat Number	Solution Volume (mL)	Wet Roll (min)	Dry Time (min)	Drying Air Temp °C
	29	250	1.0	3.0	65
	30	250	1.0	3.0	65
	31	250	2.0	3.0	65
	32	250	2.0	3.0	65
	33	250	2.0	3.0	65
	34	250	2.0	3.0	65
	35	250	2.0	3.0	65
subtotal	8500 mL	40.0	105		145 total min
Color Syrup					
	1	250	1.0	3.0	55
	2	225	1.0	3.0	55
	3	200	2.0	3.0	55
	4	200	2.0	3.0	55
	5	200	2.0	3.0	55
	6	200	2.0	3.0	55
	7	175	2.0	3.0	55
	8	175	2.0	3.0	55
	9	175	2.0	3.0	55
	10	150	2.0	2.0	55
	11	150	2.0	2.0	55
	12	125	2.0	2.0	55
	13	125	2.0	2.0	55
	14	125	3.0	0	55
	15	100	3.0	0	N/A
subtotal	2575 mL	30	35		65 total min
Finish Syrup					
	1	50	3.0	0	N/A
	2	40	3.0	0	N/A
subtotal	90 mL	6	0		6 total min
TOTAL					216 minutes

prior to color coating. The high solids content of the grossing solution (80 percent) allows fast water evaporation. Thus, a seal coat may not be required unless physical deterioration of the core tablet is observed or chemical instability of the active is noted.

In this example, the initial amount of solution applied to the core tablets is steadily increased to fill the edges and round the corners. The longer wet roll times in the last five applications allow smoothing of the surface prior to color coating. The application of smoothing coats may be required if the coated surfaces are not smooth enough to accept the color coats.

Color Coats

In this example, the volume of color syrup is decreased with subsequent applications as the tablet surface becomes smoother. The wet roll times are increased and the dry times are reduced to aid the formation of smoother tablet surfaces.

Finish Coats

The two finish coat applications of 60 percent sugar syrup allow a very fine crystalline surface to develop, which is necessary for the application of polishing solutions. The tablets are tray dried overnight before the polishing operation, as discussed previously on page 32.

AUTOMATED SUGAR COATING

The following discussion considers the requirements for an automated sugar-coating process that uses a perforated pan.[4]

Process Overview

Conventional sugar-coating processes can be automated to minimize the amount of labor required, provide better process control, and improve product uniformity and quality (Anderson and Sakr 1996; Krause and Iorio 1968; Wood and Scarpone 1983). Processing is controlled automatically by using a programmable controller. The spray, mix, and dry times, along with damper and process air control, are governed by parameters entered into a programmable controller.

An automated sugar coating process consists of four steps, which constitute a cycle. Each cycle is repeated using various coating solutions until the desired effect is achieved. The steps are as follows:

---

4. Correspondence, 1997, from the Vector Corporation, Marion, IA.

- Spray: Solution is applied.

- Distribute solution: Cores are tumbled to disperse the solution. There is no air movement through the pan.

- Initial drying: Drying begins as room temperature air is pulled through the pan by the exhaust fan.

- Final drying: Drying continues until sufficient moisture is removed (Fox et. al. 1977; Heyd and Kanig 1970). Both inlet and exhaust fans are utilized to move conditioned and/or heated air.

Core Tablet Properties

Core tablet properties are similar for both conventional and automated sugar-coated processes.

Equipment

Coating Pans

In perforated pans, process air is drawn through the tumbling product bed. Perforated pans, by design, tend to be more efficient users of energy. This design increases the drying efficiency, which results in shorter coating times.

Solution Application Apparatus

Solution is applied via an air-atomized spray gun; this allows uniform coverage and eliminates localized overwetting. More uniform and quicker wetting is also achieved. Sugar solutions can be applied with both high and low pressure equipment[5] such as the following:

- Airless spray guns.

- Air atomization spray guns (see Figure 2.8).

- Pipe guns with spray tips.

- Pipe guns with holes.

Solution type and viscosity are the primary factors in determining the type of spray apparatus. Airless spray guns are normally used for high-viscosity solutions that cannot be air atomized.

---

5. Equipment is available from Binks Manufacturing Company, Franklin Park, IL; Graco Inc., Minneapolis, MN; and Spraying Systems, Wheaton, IL.

Figure 2.8. FMG air atomizing spray gun. Courtesy of Vector Corporation.

During the sugar application and mixing steps, it is very important to apply the sugar syrup as quickly and evenly as possible across the entire surface of the product bed. It is recommended that the spray apparatus have high flow rates, a wide pattern, and minimal droplet size of the solutions. The following options can be considered:

- With very high solid content solutions, it may be necessary to heat the spray manifold to prevent sugar crystallization, which leads to clogging of the spray guns and manifold.

- Solution recirculation back to the tank is normally necessary with high solid content solutions to ensure uniformity.

- Air purging spray heads can be used to clean spray guns after each spray application.

Many solutions used for sugar coating are very viscous and have a high percent of solids in suspension. Under these conditions, heated solution lines are necessary to maintain proper flow. Lines are normally heated with hot water or electric heat tape.

Solution Pump

Any sugar-coating system is dependent on its ability to pump, regulate, and evenly apply the coating solution. The selection of the feed system should be based on a number of process and application considerations, including the following:

- Solution type and viscosity.
- Jacketed system required.
- Controlling the application rate.
- Space limitations.
- Flow rate.
- Materials of construction.
- Cleanup.
- Efficiency.
- Variation in required solutions.
- Cost.

Most sugar-coating applications require the use of heated, high-viscosity sugar solutions. These solutions require a moderate degree of accuracy for the individual and total sugar applications. The most common types of pumps or feed systems used for sugar coating are as follows:

- Positive displacement piston type[6] (Figure 2.9).
- Gear or lobe type[7] (Figure 2.10).
- Moyno[8] (Figure 2.11).
- Pressurized vessels[9].
- Diaphragm[10].

When evaluating the process and application considerations, the positive displacement lobe type pump is recommended because of its flexibility, accuracy, and cost. Lobe type pumps may be utilized with any normal sugar-coating solution. A measured amount of so-

---

6. Graco Inc., Minneapolis, MN.
7. Waukesha Cherry-Burrell, Delavan, WI.
8. Moyno Industrial Products, Springfield, OH.
9. Graco Inc., Minneapolis, MN.
10. Cole-Parmer Instrument Company, Vernon Hills, IL.

Figure 2.9. Positive displacement piston pump. Courtesy of Graco Inc.

lution is delivered with each revolution. Spray initiation can be either manual or automated. The pump size should be determined by batch size; all system parameters, such as the maximum application per step, can be achieved in approximately 10 to 15 sec.

Process Example

The following automated sugar-coating illustration utilizes a Hi-Coater® Model H-100 (39 in. pan) fitted with four flush mount baffles. See Figure 2.12 for representative Hi-Coater® automated

Figure 2.10. Waukesha gear pump, interior view. Provided courtesy of Schering-Plough HealthCare Products.

Figure 2.11. Moyno pump, interior view. Courtesy of Moyno Industrial Products.

Figure 2.12. Automated coating system schematic. Courtesy of Vector Corporation.

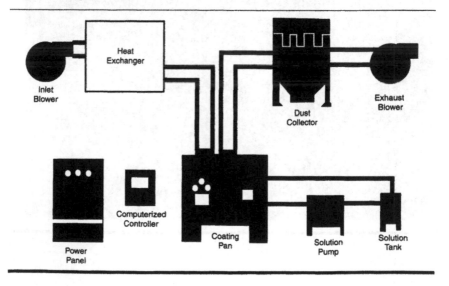

sugar-coating process schematics and Figure 2.13 for coating sequence schematics. The tablet load, formulas, and process parameters are listed in Table 2.12. This example uses medicated chewing gum as the core to be coated. The coating process is similar to that for a conventional tablet core.

Subcoating

A 76 percent sugar with 1 percent starch solution was used for the subcoating. The cores were not sealed since they would not be subjected to a high moisture level like that in a conventional process. However, should a seal coat be necessary, it could be applied in a fashion similar to that of film coating, eliminating the need to dry charge dusting powder.

Examining the coating process parameters in Table 2.13 for this trial shows that the amount of solution applied to the cores was steadily increased to fill the edges and round the corners. Once rounding was complete, the solution volume was reduced to generate a smooth surface.

Finish Coating (Clear Coat)

The last three applications of clear coat solution in Table 2.13 consisted of spraying the surface with a 67 percent sugar solution to prepare the surface for polishing, as in conventional sugar coating. This automated coating example illustrates the possibility of eliminating some processing steps (i.e., seal coating and subcoating), minimizing operator variation, and reducing process time from a 2–3-day conventional operation to a 1-day automated process.

QUALITY CONTROL TESTING

Quality control testing of sugar-coated tablets includes the chemical and physical attributes of the coated product.

Chemical Testing

Assay and Content Uniformity

A chemical assay substantiates conformance of the product active content to a monograph or the manufacturer's requirements.

Figure 2.13. Automated coating sequence schematic. Courtesy of Vector Corporation.

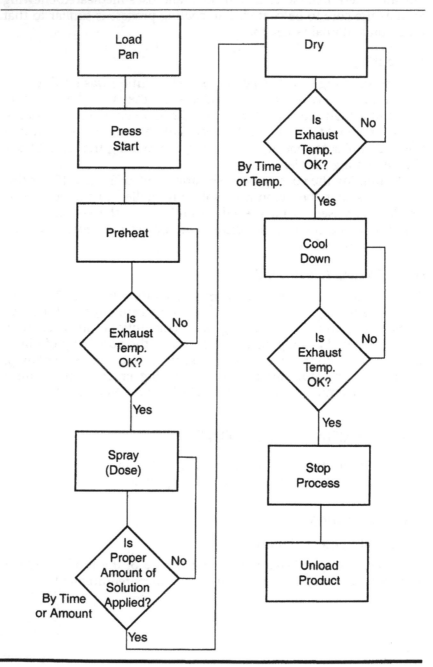

**Table 2.12. Automated Sugar Coating Process Example[a]:
Equipment/Parameters**

Medicated Chewing Gum—Perforated Pan	
Tablet:	Medicated Chewing Gum Core
	Weight: 1.05 g
Solutions:	Subcoat: 70% Sucrose, 1% Corn Syrup
	Clear Coat: 67% Sucrose
Coating Conditions:	
	Hi-Coater Model HC-100
	Loading Volume 30.0 kg
	Pan Rotation Speed 12 rpm
	Sugar Pump Waukesha w/PC
	Spray Gun Model FMG
	Spray Pressure 18 psi
	Spray Distance 10 in.
	Ambient Relative Humidity 39%
	Drying Air Temperature 75°C
	Solution Temperature 70°C
Results:	
	Coating Process Time 123 min
	Amount of Solution Used 9.375 L subcoat
	0.375 L clear
	Process Tablet Weight 1.75 g
	Weight Increase 25.8% (theoretical)

a. Correspondence, 1997, from Vector Corporation, Marion, IA.

Physical Testing

Physical testing may include disintegration time, dissolution, dimensional specifications, and visual attributes of the sugar-coated product.

Table 2.13. Automated Sugar Coating Process Example—Data[a]

Medicated Chewing Gum—Perforated Pan

	SPRAY (mL)	PAUSE (min)	Room Air Exhaust	Exhaust Dry (min)	Total Time (min)	Solution Composition
Subcoat						
1	169	2.0	0.5	1.5		76% sugar/
2	169	2.0	0.5	1.5		1% starch
3	169	2.0	0.5	1.5		
4	260	2.0	0.5	1.5		
5	260	2.0	0.5	1.5		
6	260	2.0	0.5	1.5		
7	343	2.0	0.5	1.5		
8	343	2.0	0.5	1.5		
9	343	2.0	0.5	1.5		
10	343	2.0	0.5	1.5		
11	343	2.0	0.5	1.5		
12	343	2.0	0.5	1.5		
13	343	2.0	0.5	1.5		
14	343	2.0	0.5	1.5		
15	450	2.0	0.5	1.5		
16	450	2.0	0.5	1.5		
17	450	2.0	0.5	1.5		
18	525	2.0	0.5	1.5		
19	525	2.0	0.5	1.5		
20	525	2.0	0.5	1.5		
21	525	2.0	0.5	1.5		
22	450	2.0	0.5	1.5		
23	450	2.0	0.5	1.5		
24	343	2.0	0.5	1.5		
25	343	2.0	0.5	1.5		
26	343	2.0	0.5	1.5		
27	343	2.0	0.5	1.5		
Total	9.753 L	54.0	13.5	40.5	108.0	
Finish Coat						
28	125	3.0	0.5	1.5		67% Sucrose
29	125	3.0	0.5	1.5		
30	125	3.0	0.5	1.5		
Total	0.375 L	9.0	1.5	4.5	15.0	
					123.0	

a. Correspondence, 1997, from Vector Corporation, Marion, IA

Disintegration Time

Sugar coating adds to the overall disintegration time of a core tablet, especially if a seal coat has been applied. Therefore, disintegration of the sugar-coated product within a specified time—whether or not a seal coat is present—is a critical parameter to be monitored.

Dissolution

Dissolution may be required in some cases. For example, an immediate release, sugar-coated product or an enteric sugar-coated product may have a monograph dissolution specification that must be met.

Dimensional Specifications

Thickness is critical for a sugar-coated product to be branded and eventually packaged in either bottles or blister cards. Equipment for these operations may utilize slats with product cavities to contain and transfer single- or multiple-coated tablets and molds for forming pockets in blister cards to hold the product. Therefore, the product must be within certain dimensional tolerances to preclude problems with these operations. Thickness is normally monitored during the coating process to limit the product within a certain range. After coating, sorting is performed to remove any oversized and or undersized coated product units.

Visual Inspection

Random sampling of the finished product is conducted to ascertain that the visual appearance of the product meets internal product quality levels. The defects that should be monitored include rough coating, mottled color, chipping, and improper branding.

REFERENCES

Anderson, W., and A. M. Sakr. 1996. Coating of pharmaceutical tablets: The spray-pan method. *J. Pharm. Pharmacol.* 18:783–794.

Burroughs, J. 1996. Gelatin solutions. *The Manufacturing Confectioner* (October): 35–38.

Fox, D. C., A. E. Buckpitt, M. V. Laramie, and M. E. Miserany. 1977. Automated process for sugar coating tablets. Paper presented at A.I.Ch.E. Meeting, 17 November, in New York.

Hayes, J. T. 1996. *Pharmaceutical solid dosage product printing.* Moorestown, N.J., USA: Ackley Machine Corporation.

Heyd, A., and J. L. Kanig. 1970. Improved self-programming automated tablet-coating system. *J. Pharm. Sci.* 59:1171–1174.

Khalil, S. A. H., N. S. Barakat, and N. A. Boraie. 1991. Effect of ageing on dissolution rates and bioavailability of riboflavine in sugarcoated tablets. *S.T.P. Pharma Sciences* 1 (3):189–194.

Krause, G. M., and T. L. Iorio. 1968. Application of sugar coating to tablets and confections by means of an automated airless system. I. *J. Pharm. Sci.* 57:1223–1227.

Lehman, K., and B. Brögmann. 1996. Tablet coating. In *Encyclopedia of pharmaceutical technology,* edited by J. Swarbrick and J. C. Boylan. New York: Marcel Dekker, Inc., vol. 14, pp. 355–384.

Mantrose-Bradshaw–Zinnser Group. *Pharmaceutical Glazes, NF.* Technical Bulletin No. 70-09-01.

Porter, S. C. 1981. Tablet coating. *D&CI* (May): 46–93.

Robinson, M. J. 1980. Coating of pharmaceutical dosage forms. In *Remington's pharmaceutical sciences,* 16th ed., edited by A. Osol, G. D. Chase, A. R. Gennaro, M. R. Gibson, C. B. Granberg, S. C. Harvey, R. E. King, A. N. Martin, E. A. Swinyard, and G. L. Zink. Easton, Penn., USA: Mack Publishing Company.

Seitz, J. A., S. P. Mehta, and J. L. Yeager. 1986. Tablet coating. In *The theory and practice of industrial pharmacy,* 3rd ed., edited by L. Lachman, H. A. Liebermann, and J. L. Kanig. Philadelphia: Lea & Febiger, pp. 346–373.

Sonndecker, G., and G. Griffenhagen. 1990. Sugarcoating the pill: A historical perspective. *Pharm. Tech.* (September): 77–80.

Wood, T. G., and A. J. Scarpone. 1983. Control and automation of the sugar coating of tablets. In *Automation of pharmaceutical operations,* edited by D.J. Fraade. Springfield, Ore., USA: Pharmaceutical Technology Publications, pp. 257–262.

APPENDIX: SUPPLIERS

Ackley Machine Corporation
1273 North Church Street
Moorestown, New Jersey 08057
Phone: 609-234-3626

Glatt Air Techniques
20 Spear Road
Ramsey, New Jersey 07446
Phone: 201-825-870

Graco Inc.
P.O. Box 1441
Minneapolis, Minnesota 55440
Phone: 800-731-3926

Moyno Industrial Products
P.O. Box 960
Springfield, Ohio 45501
Phone: 937-327-3111

Nicomac, Inc. (USA)
10 W. Forest Avenue
Englewood, New Jersey 07631
Phone: 201-871-0916

Thomas Engineering Inc.
P.O. Box 198
575 West Central Road
Hoffman Estates, IL 60195
Phone: 847-358-5800

Vector Corporation
675 44th Street
Marion, Iowa 52302
Phone: 319-377-8283

Schering-Plough HealthCare Products
3030 Jackson Avenue
Memphis, Tennessee 38151
Phone: 901-320-4915

3

FILM COATING

Robert J. Campbell

Chemical Engineering & Instrumentation Consultants Inc.

Gary L. Sackett

Vector Corporation

Tablet coating is one of the oldest pharmaceutical arts still in existence, dating back to around 850 A.D. The film coating processes of today originated in the early 1950s with the advent of new types of raw materials. The commercial use of the fluid bed coater began in the late 1950s with the introduction of the Wurster Air Suspension Process. Aqueous film coating was introduced in the early 1970s with the invention of highly efficient panning equipment. Today, aqueous film coating has truly become a science with the recent development of sophisticated coating polymers and automated coating systems.

There are a number of benefits that can be derived from film coating:

- Mask unpleasant odors or tastes.

- Improve the ease of ingestion.

- Improve the product's appearance.

- Protect the product from its surrounding environment (i.e., air, moisture, or light).

- Control the rate of release (i.e., enteric or sustained release).

- Separate incompatible materials.

- Improve the product's identification.

- Facilitate handling.

- Prevent contaminants from reacting with the product.

- Apply an active coating to a substrate (i.e., apply active as a solution or a slurry).

THEORY

Atomization/Coalescence/Drying

Simply stated, film coating is the application of a film-forming polymer onto a product substrate. The challenge of film coating is to apply the spray droplets uniformly and for all of the droplets to dry at the proper evaporative rate. The spray droplets must contain the proper level of liquid when they strike the tablet surface. If the spray droplets contain too much liquid, then overwetting of the substrate will occur. If the droplets are too dry, then they will not spread out or coalesce to form a smooth film. A number of variables (sometimes seeming almost infinite) can cause difficulties in the pursuit of an acceptable coating.

Weight Gain

Film coating may be applied until the tablets reach a targeted weight range. Alternatively, the quantity of film coating may be expressed as a percentage weight gain, which can be calculated as illustrated in the following example:

- Uncoated core weight = 250 mg.

- Desired weight gain = 3 percent.

- Quantity of film coating per tablet = 250 mg × 0.03 = 7.5 mg.

A typical weight gain for an aesthetic coating is 2–3 percent; for clear coatings, the weight gain may be as little as 0.5–1.0 percent. For controlled-release applications, the quality of the film is directly related to its thickness. A thickness of 30–50 μm (i.e., 4–6 mg polymer/cm^2 of tablet surface) is usually sufficient to provide a satisfactory enteric coating (Röhm GmbH 1996a). Since a batch of smaller tablets contains a greater total tablet surface area, it will require a greater weight gain to achieve a controlled-release film of a suitable thickness.

APPLICATIONS

There are several different applications (methods) for the coating of pharmaceutical dosage forms. The reasons for selecting film coating are essentially the same whether using a perforated coating pan or the fluid bed coating process. In general, the major criterion for deciding between whether to use a coating pan or fluid bed coating is the size of the product substrate. A rule of thumb is that if the product is approximately 1/4 in. or less in diameter, the preferred equipment for coating is the fluid bed coater. When coating small particles in a perforated coating pan (with airflow that is drawn down through the tablet bed), the pressure drop across the product bed drastically reduces the process air volume, a factor that will reduce the evaporative capacity of the pan. Coupled with the close contact of the product in the spray zone, the coating of small particles in the coating pan becomes less desirable. For the coating of large product substrates, the relatively rough product movement experienced in the fluid bed makes it less appealing. Additionally, it may be difficult to achieve an acceptable product movement with large, irregular-shaped tablets in a fluidized bed coater.

RAW MATERIALS

Sources of raw materials are listed in the appendix to this chapter.

Film Polymers

As one would expect, the film coating polymer is the most critical component in the coating solution. Both water-soluble and -insoluble film polymers are commonly used for coating. Some of the key attributes that the film polymer must possess are as follows:

- Continuous film formation capability.
- Compatibility with the product substrate.
- Low viscosity in order to atomize adequately.
- Solubility in the desired solvent.

These polymers generally require the addition of a plasticizer to reduce the brittleness of the resultant film. Some of the more common film polymers are listed in Tables 3.1, 3.2, and 3.3.

Water-Soluble Polymers

Water-soluble coatings are usually added to improve the appearance of the product or to mask unpleasant tastes. However, they may also provide protection from light, air, physical stress, and so on. The characteristics of various water-soluble polymers are listed in Table 3.1.

Water-Insoluble Polymers

Delayed-release coatings typically fall into two categories: sustained release or enteric release. Examples of these polymers are listed in Tables 3.2 and 3.3, respectively.

Plasticizers

Without the addition of a key coating solution excipient, film coatings may be prone to cracking. This key excipient is a plasticizer, and its function is to reduce the brittleness of the polymer. The plasticizer reduces the glass transition temperature (T_g) of the coating, which allows the film to remain flexible at room temperatures. If excess plasticizer is added, then the film may be sticky and agglomeration of the product may occur during storage. Both water-soluble and water-insoluble plasticizers are commonly used. Effective plasticizers are generally soluble in the polymer they plasticize (Seitz et al. 1986). The plasticizer must be miscible and interact with the polymer. Typically, water-soluble plasticizers are used with water-soluble film polymers, and water-insoluble plasticizers are used with water-insoluble film polymers. However, water-insoluble plasticizers can be emulsified for use with aqueous acrylic dispersions (Morflex). Film-forming polymer suppliers will usually have recommendations for the preferred plasticizer and usage level. Some of the more common plasticizers are as follows:

Water-Soluble Plasticizers

Polyethylene Glycol (PEG). This is the most commonly used plasticizer. It is available in a variety of different molecular weights (ranging from 300 to 20,000). PEG (available as Carbowax®) with a molecular weight of 400 is in liquid form at room temperature and

Table 3.1. Water Soluble Polymers

Polymer	Key Physical Properties	Processability	pH Solubility	Other
Hydroxypropyl methylcellulose (HPMC)	Soluble in water and a wide range of organic solvents	Dispersed with high agitation into solvent or heated water	Soluble in either acidic or basic fluids	Easy to use. Most commonly used polymer for immediate release
Methylcellulose (MC)	Soluble in water and a wide range of organic solvents	Similar to that of HPMC	Soluble in either acidic or basic fluids	Films have a higher gel strength and a lower gel temperature than HPMC films (Dow Chemical 1996).
Hydroxypropyl cellulose (HPC)	Soluble in water and a wide range of organic solvents	Dispersed with high agitation into solvent or water; insoluble in water above 45°C (Hercules 1984)	Soluble in either acidic or basic fluids	Produces a film that is slightly more tacky than those with HPMC

Continued on next page.

Continued from previous page.

Polymer	Key Physical Properties	Processability	pH Solubility	Other
Polyvinyl-pyrrolidone (PVP)	Readily soluble in water and a variety of organic solvents	Produces a tacky spray during coating. Film may become sticky after application due to the hygroscopic nature of the polymer (GAF 1981).	pH independent	More commonly used as a binder for granulation processes
Food Starch—Modified	Insoluble in cold water and a variety of organic solvents	Some combinations of polymers and plasticizers may produce coatings that are slightly tacky (Grain Processing Corp. 1995).	pH independent	HPMC can be added to reduce the tackiness of the coating. Starch films do not require the addition of a plasticizers. Available as PURE-COTE® from Grain Processing Corp.

Table 3.2. Sustained-Release Polymers

Polymer	Key Physical Properties	Processability	pH Solubility	Other
Ethylcellulose (EC)	Produces a diffusion-controlled membrane. Soluble in most organic solvents and inert to most alkalis.	Can mix with water soluble polymers to moderate the release	pH independent	Available as aqueous dispersions: Surelease® (Colorcon) or Aquacoat® (FMC)
Methacrylic acid copolymers	Eudragit® RL/RS produces a delayed release action.	Aqueous dispersions are thermoplastic, hence low product temperature must be maintained. Additives can be used to reduce tendency for product agglomeration/twinning (Röhm GmbH 1996b).	The permeability of Eudragit® RL and RS films is independent of pH.	Available as Eudragit® from Rohm Tech Inc.

Table 3.3. Enteric-Release Polymers

Polymer	Key Physical Properties	Processability	pH Solubility	Other
Hydroxypropyl methylcellulose phthalate (HPMCP)	Preferred organic solvent mixtures are chloride/alcohol or acetone/alcohol.	Does not always require a plasticizer (Shinetsu Chemical 1974).	HPMCP will dissolve in an alkali or in USP buffer with a pH of 5.0 or higher.	Available from Shinetsu Chemical Co. or Eastman Fine Chemicals
Cellulose acetate phthalate (CAP)	Soluble in a number of ketones, esters, ethyl alcohol, and cyclic esters.	Aqueous solutions can be produced by adding ammonium hydroxide. Although some additives are sensitive to ammoniated polymer.	Soluble in USP buffer solutions with a pH ≥ 6.2.	Introduced in the 1940s by Eastman Kodak Co. Most commonly used enteric polymer. Available as aqueous dispersion: Aquacoat®CPD (FMC).
Polyvinyl acetate phthalate (PVAP)	Soluble in a number of organic mixtures and solvent mixtures.	Excellent moisture barrier, also used as a tablet sealant.	Soluble in USP buffer solutions with pH ≥ 5.0.	Available as aqueous dispersion: Sureteric® (Colorcon).

Continued on next page.

Continued from previous page.

Polymer	Key Physical Properties	Processability	pH Solubility	Other
Methacrylic acid copolymers	Eudragit®L/S provides enteric release.	Aqueous dispersions are thermoplastic, hence low product temperature must be maintained. Additives can be used to reduce the tendency for product agglomeration/twinning.	Eudragit® L is soluble in buffered solutions of pH ≥ 5.5, while Eudragit® S is soluble at pH ≥ 7.0.	Available as a powder, organic solvent solution, or aqueous dispersion.

is used as a plasticizer for the coating of small particles. Higher molecular weight forms of PEG (3,350 or greater) are customarily used with tablet coating formulations.

Propylene Glycol (PG). Normally used as a plasticizer for small particle coatings, it may also be used for tablet coating applications. It produces a softer, more flexible film than higher molecular weight forms of PEG.

Others. The following substances can also be used as water-soluble plasticizers:

- Triethyl citrate (TEC): water soluble to 6.5 percent (available as Citroflex®).
- Triacetin.
- Glycerin.

Water-Insoluble Plasticizers

The following substances are water-insoluble plasticizers:

- Tributyl citrate (TBC).
- Acetylated monoglyceride (AMG).
- Castor oil.
- Dibutyl sebacate (DBS).
- Acetyl triethyl citrate (ATEC).
- Acetyl tributyl citrate (ATBC).

Colorants

Colorants are used to impart a color to the film coating. Dyes are water-soluble colorants; pigments or lake colors are water-insoluble colorants. Color development in film coating is much more efficient with an opaque suspension; therefore, use of an aluminum lake color is recommended. An aluminum lake color is produced through the absorption of a dye molecule onto an alumina substrate (Colorcon Inc.). Aluminum lake colors are available as dispersible dry powders or as concentrated suspensions. Some of the prepared color concentrates are as follows:

- Opaspray®: A product of Colorcon, Inc., Opaspray® is a concentrated suspension of lakes, pigments, and other additives in 3-A alcohol. It is manufactured to a solids level of approximately 40 percent.

- Chroma-Tone®: A dispersible dry color concentrate available from Crompton & Knowles Corporation.

- Chroma-Kote®: Available as a concentrated pigment dispersion in an aqueous or PG base from Crompton & Knowles Corporation.

- Spectraspray®: A product of Warner Jenkinson Company, Inc., Spectraspray® is available as an aqueous color concentrate. Typical colorant solids levels vary from 30 to 50 percent.

Organic Solvents

The use of an aqueous-based film coating has steadily replaced coatings with organic solvents. This decline was the result of the introduction of more efficient perforated coating pans in the early 1970s. However, certain water-insoluble film polymers still have a niche for organic solvent usage. Concerns with environmental issues, coating economics, and operator safety have made it desirable, whenever possible, to switch to aqueous coating. The volatility of organic solvents did offer a major advantage in the increased rate at which film coatings could be applied. At one time, methylene chloride/methanol was the preferred solvent combination due to its reduced flammability risk; however, toxicity hazards associated with methylene chloride have reduced its appeal. The following are some of the more common solvent/solvent combinations:

- Ethanol/water.

- Acetone/water.

- Methanol.

- Methylene chloride/methanol.

- Methylene chloride/ethanol.

- Acetone/methylene chloride.

- Acetone/ethanol/isopropanol.

- Acetone/ethanol/methylene chloride.

Preformulated Coating Blends

Preformulated dry blends contain one or more of the following materials: polymer, plasticizer, and pigment. All that is required is the addition of water. These materials exhibit improved ease of hydration and can be easily prepared in about 30 minutes. Some examples include the following:

- Opadry I®/Opadry II®: Available from Colorcon, Inc., these coating blends are prepared to a solids levels of 15–20 percent. They contain polymer, plasticizer, and pigment.

- Dri-Klear®: Available from Crompton & Knowles Corporation, Dri-Klear® is a combination of a cellulosic polymer and a plasticizer. The recommended usage level is 10.0 percent; however, a range of 5.0 to 14.0 percent may be used (Crompton & Knowles 1992).

FILM COATING FORMULATIONS

General guidelines for solution preparation are to avoid excessive agitation and to filter the coating solution prior to use. Excessive agitation during solution preparation will result in air entrapment, which can lead to inaccuracies in solution delivery or create delays due to the time required to deaerate the solution. Furthermore, some aqueous dispersions are shear sensitive, and precipitation of the polymer may occur with vigorous agitation of the solution. Filtration of the solution is strongly recommended to prevent the plugging of the spray gun. The solution can either be screened after preparation, or an in-line solution filter can be used.

Guidelines/Limitations

Film Mechanical Strength

The film formulation must be prepared so that the resultant film has adequate mechanical strength. The reasons for this are twofold:

1. The film must be strong enough to protect the tablet from excessive attrition due to tumbling.

2. The film must be able to resist the erosion of the film itself.

A weak film will usually exhibit wear or erosion at the tablet edges. The addition of too many nonfilm-forming excipients (e.g., mineral-type fillers) will decrease the strength of the film. This may occur if a drug is added to the coating formulation at a high level or if high levels of insoluble colors or mineral fillers are added.

Plasticizer Level

The function of a plasticizer is to reduce the T_g of the film (i.e., to produce a film that is not brittle at normal operating conditions). If the film is inadequately plasticized, then it will be too brittle and will be more prone to cracking. On the other hand, if a film solution has an excessive level of plasticizer, then the mechanical strength of the film may be reduced.

Pigment Level

If the coating formulation contains insufficient pigment, it will be impossible to develop the desired color intensity. Additionally, low pigment levels can make it difficult to minimize color variation due to poor opacity of the coating formulation. High pigment levels can, as stated earlier, reduce the mechanical strength of the coating.

Solids in Film Formulations

A low solids level in the coating formulation can increase the process time and the time required to provide a protective film coating resulting in increased tablet attrition. High solids levels coupled with low coating can make it difficult to achieve acceptable film coating uniformity. High solids levels can also be a problem if it results in a high viscosity. If a high solids level does not create an excessively viscous formulation, it provides an excellent opportunity to reduce the volume of coating formulation and, thus, the coating time. The aqueous dispersions available on the market have an extremely low viscosity.

Formulation Viscosity

Formulation viscosity is closely tied to the concentration of solids. As the viscosity increases (above ~200–250 cps), the droplet size distribution produced by the spray guns becomes increasingly wider due to the production of larger droplets. Slight increases in viscosity can be compensated for by increasing the volume of atomization air. However, at viscosities greater than ~350 cps, it is difficult to eliminate all of the larger droplets. To compensate for the presence of larger droplets, the operator will usually increase the inlet temperature to prevent these larger droplets from overwetting the product. This will lengthen the amount of spray drying due to the premature evaporation of water from the finer droplets. The net result is a lower coating efficiency and a rougher film surface.

Stability

The coating formulation must be stable (both physically and chemically) for the duration of the coating time. The most typical instability problem encountered is the settling out of solids. This occurs with formulations that contain an excessive percentage of solids or when the coating has insufficient suspending capacity. The settling of solids can also lead to a blockage of the solution lines or the spray guns.

Compatibility

Compatibility applies to the coating materials and between the film coating and the tablet core. For example, the addition of color concentrate that contained ethanol has led to the precipitation of some of the polymers used in aqueous dispersions. As a remedy, a PG–based color was used. In another product, there was an interaction between the plasticizer used and the active used in the tablet core. It was necessary to reformulate with an alternative plasticizer (Sackett 1997).

Types of Coatings

Taste Masking/Aesthetic Coatings

Customarily, coatings used for taste masking or aesthetic appearances are immediate release films. If a delayed-release polymer is used, it is blended with a readily water-soluble material to provide a rapid release of the active.

- Typical film coating compositions:

Film-forming polymer	7.0–18.0 percent.
Plasticizer	0.5–2.0 percent.
Pigment/colorant	0.0–8.0 percent.
Solvent: up to	100.0 percent.

- Sample formulation:

HPMC (5 cps)	8.0 percent.
PEG 3350	2.0 percent.
Colorant	5.0 percent.
Deionized water	85.0 percent.
	100.0 percent.

- Solution preparation:

    1. Heat 1/3 to 1/2 of the required water to 80°C.

    2. Using moderate agitation, add the HPMC to the vortex. Mix for 5 to 15 min.

    3. Using mild agitation, add the remaining water (cold, 5–15°C).

    4. When deaeration of the solution is complete (this may require 1–4 h depending on the speed of agitation), add the remaining coating materials and mix thoroughly.

Barrier Coatings

Applications. Barrier coatings are used to protect the product from air, moisture, or light hazards. They can also be used as a moisture barrier prior to the application of sugar coatings. In some cases, solution formulations used for aesthetic coatings can be used to produce a barrier film. One example of this is the application of an organic solvent–based HPMC coating on an effervescent tablet. Solutions ready for use as barrier coatings, such as zein, a natural corn protein, or Opaseal®, an alcoholic solution of PVAP, are available from Colorcon, Inc..

- Sample formulations:

    Example 1:

HPMC (5 cps)	6.0 percent.
Shellac	4.0 percent.
Ethanol	40.0 percent.
Methylene chloride	50.0 percent.
	100.0 percent

    Example 2:

HPMC (15 cps)	5.0 percent.
EC (7 cps)	1.0 percent.
Acetylated monoglycerides	1.0 percent.
Ethanol	39.0 percent.
Methylene chloride	54.0 percent.
	100.0 percent

Example 3:

Aquacoat® (30 percent w/w)	21.5 percent.
HPMC (5 cps)	6.7 percent.
DBS	1.3 percent.
Opaspray®	7.0 percent.
Water	63.5 percent.
	100.0 percent.

Active Coatings

Applications. Coatings can be used to add an active compound (or an additional active) to a product substrate. The active can be applied in a number of ways: an active solution, a slurry containing an active, or a fine powder in conjunction with a binder solution spray.

- Sample formulation:

Eudragit® RS 30 D	12.50 percent.
Active substance	12.50 percent.
Talc	3.75 percent.
TEC	0.75 percent.
Water	70.50 percent.
	100.00 percent.

Enteric Coatings

Applications. Enteric coatings resist breakdown in the stomach, but they dissolve rapidly in the small intestine. This is beneficial for materials that are preferentially absorbed in the small intestine or are inactivated or cause irritation if released in the stomach.

Examples.
- Organic solvent–based formulations

Example 1:

CAP	5.0 percent.
Diethyl phthalate	2.0 percent.
Lake color	0.25 percent.

Titanium dioxide	0.50 percent.
Talc	0.25 percent.
Ethyl acetate/isopropanol (1:2 w/w)	92.0 percent.
	100.0 percent.

Example 2:

HPMCP HP-55	8.0 percent.
Methylene chloride	46.0 percent.
Methanol	46.0 percent.
	100.0 percent.

Aqueous dispersions:

Example 1:

Eudragit® L 30 D	50.0 percent.
TEC	1.5 percent.
Talc	7.5 percent.
Simethicone emulsion	0.1 percent.
Water	40.9 percent.
	100.0 percent.

Example 2:

Sureteric®	15.00 percent.
Antifoam emulsion (Dow Corning)	0.05 percent.
Water	84.95 percent.
	100.00 percent.

Suspension preparation:

1. Add the antifoam emulsion to the required quantity of water.

2. With moderate agitation, add the Sureteric® powder to the water. Maintain a vortex throughout the powder addition.

3. After the Sureteric® has been added, reduce the mixer speed to slow and mix for an additional 30–45 min.

4. Screen the suspension through a 60-mesh sieve prior to use.

5. Maintain a gentle agitation of the suspension during the coating process.

Example 3:

Aquateric®	11.3 percent.
TEC	3.4 percent.
Tween® 80	0.3 percent.
Water	85.0 percent.
	100.0 percent.

Suspension preparation:

1. Dissolve Tween® 80 in the required quantity of water.

2. Add TEC and mix until a clear solution or emulsion is formed.

3. Add the Aquateric® powder slowly using a propeller-type stirrer and moderate agitation. Mix for 90 min; no heating or cooling is required.

4. Maintain moderate agitation of the dispersion during coating.

- Aqueous solutions:

HPMCP	9.6 percent.
Myvacet® 9-45	2.4 percent.
Ammonium hydroxide	1.7 percent.
Deionized water	86.3 percent.
	100.0 percent.

Solution preparation:

1. Disperse HPMCP in water with stirring.

2. Add Myvacet® 9-45.

3. Add ammonium hydroxide.

4. The solution should turn clear in about 45 min.

Sustained-Release Coatings

Applications. Sustained release coatings, as the name implies, slowly release their contents over an extended time period. These coatings provide a more uniform and effective release of drug into the bloodstream.

Examples.

- Organic solvent–based formulations:

    Example 1:

EC	6.0 percent.
Isopropanol	56.0 percent.
Ethanol	38.0 percent.
	100.0 percent.

    Example 2:

EC	6.0 percent.
Diethyl phthalate	0.6 percent.
Opaspray®	5.4 percent.
Methanol	33.0 percent.
Methylene chloride	55.0 percent.
	100.0 percent.

- Aqueous dispersions:

    Example 1:

Surelease® (25 percent solids dispersion)	60 percent w/w.
Distilled water	40 percent w/w.
	100 percent w/w.

    Example 2:

Eudragit® RL 30 D	8.3 percent.
Eudragit® RS 30 D	25.0 percent.
Talc	8.3 percent.
TEC	2.0 percent.
Water	56.4 percent.
	100.0 percent.

Example 3:

Aquacoat® (30 percent solids)	80.7 percent.
DBS or Myvacet® 9-40	5.8 percent.
Water	13.5 percent.
	100.0 percent.

Dispersion preparation:

1. Add the required quantity of water and plasticizer to the Aquacoat®.

2. Mix for 30 min using moderate agitation.

- Hot melts: The successful application of a hot melt requires some modifications to the process equipment. The solution lines and spray gun should be heated and insulated. The air must be heated to prevent premature congealing of the hot melt material. Additionally, the hot melt should be heated and maintained at a temperature of 30 to 60°C above the melting point. The following hot melt materials can be used to produce a sustained-release coating:

  Stearic acid (melting point = 69–70°C)

  Partially hydrogenated cottonseed oil (melting point = 64–67°C)

  Hydrogenated castor oil (melting point = 86–88°C)

COATING EQUIPMENT

Equipment that can be used to apply the coatings, described in the previous section, will be discussed here. Sources for coating equipment are listed in the appendix to this chapter. The reasons for using different types of equipment will also be presented.

Batch Systems

Batch coating requires a container in which the product to be coated is placed and then heated. Figures 3.1 to 3.5 illustrate batch coating pans from several manufacturers and the different configurations available. Typically, the container will be cylindrical and rotate on a horizontal axis. A coating material will be sprayed onto the product using some type of dispersion device, typically a spray

Figure 3.1. Batch coating pan. Courtesy of Driam USA, Inc.

gun. Droplets from the spray gun are composed of a suspension that is a mixture of a liquid carrier and solids. The principle of this type of coating is to apply the suspension in such a manner that a droplet from the spray gun will land on the product and spread. The temperature of the product will then quickly evaporate the liquid from the suspension, leaving only the dried solids. The suspension viscosity must be such that any evaporation that occurs between the spray gun and the product will allow the droplet to spread before the carrier evaporates. This will leave a dry coating, composed of the solids that were in the suspension, on the surface of the product. In order to produce a homogeneous product, each tablet must have an equal probability of passing through the spray.

Continuous Systems

A continuous coating system differs from the batch system in that the tablets, once they are heated, are coated with the total amount of coating material in one pass through the spray. The systems and

**Figure 3.2. Batch coating pan.
Courtesy of Glatt Air Techniques, Inc.**

Figure 3.3. Batch coating pan.
Courtesy of Thomas Engineering, Inc.

equipment for the two types of coatings are very similar, except for the operation of the coating pan unit.

Components

The peripheral equipment for a coating system for either the batch system or the continuous coating system will be very similar. Figures 3.6 and 3.7 show equipment components typically used with batch and continuous coating systems. Some systems may not require all of these components, depending on the source for the process air.

Figure 3.4. Batch coating pan.
Courtesy of O'Hara Technologies, Inc.

System Overview

Process air is drawn in through an air handler where it will be heated, dehumidified (or humidified), and filtered before going to the coating pan (drum). The process air goes through a perforated coating pan and is exhausted through ductwork to a dust collector and a blower. Finally, the exhausted air is returned to the source for the process air. Solvent systems will require a solvent recovery unit

Figure 3.5. Batch coating pan. Courtesy of Vector Coporation.

or an incinerator before the process air can be exhausted back to the source of the process air. Some solvent systems may be a closed loop system, as shown in Figure 3.8, where the process air is recycled back to the air handler. This type of system must be designed for some loss of air after the solvent recovery unit, because compressed air will be added to the process air if pneumatic spray guns are used. A closed loop system has the advantage of using process air that is conditioned for the operating parameters; it requires less energy to operate the system, but it is more difficult to make operational.

Pan Units

Coating pans are manufactured in a variety of styles. Each can perform coating when used properly. The coating pans referenced in this section are closed operation pans (the opening to the pan is closed during operation). Coating can also be performed in open pans but at a loss of efficiency. Except for continuous coating

Figure 3.6. Typical batch coating pan system. Courtesy of Vector Corporation.

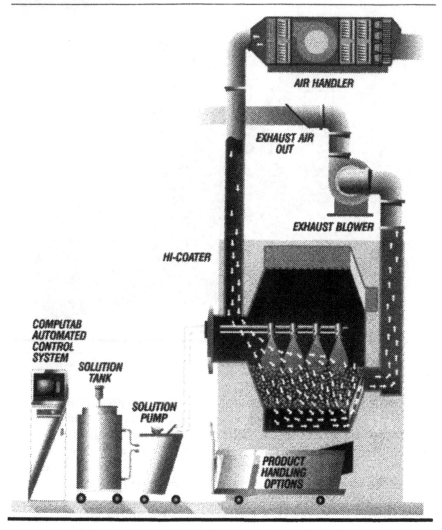

equipment, all other pans are cylindrical drums. The function of the coating drum is to provide a space where the product can reside and be moved in such a manner that all surfaces of the product are exposed for an equal amount of time to the spray. This is accomplished by rotating the drum on its horizontal axis. The drums have

Figure 3.7. Typical continuous coating pan system. Courtesy of Vector Corporation.

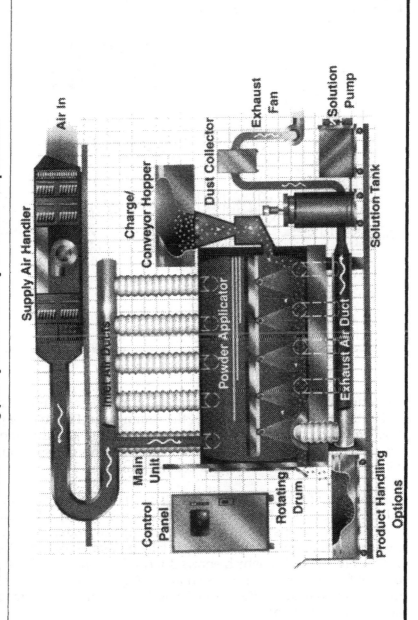

Figure 3.8. Process for closed loop coating pan system. Courtesy of Vector Corporation.

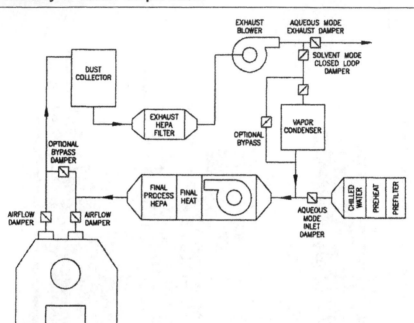

a mixing baffle that forces the product to tumble and move back and forth on the cylindrical wall of the coating pan. The coating pan must provide a means for conditioned process air to pass through or over the product for heat transfer. A means for mounting spray guns must also be provided inside the coating pan.

Comparison of Batch Type Coating Pans. The coating pan can be either fully or partially perforated around the cylindrical surface, which allows process air to be drawn through a product placed in the coating pan. An example of a partially perforated coating drum is shown in Figure 3.9, which shows ductwork on the outside of the drum. A fully perforated drum is shown in Figure 3.10, which has no ducts on the outside of the drum. Table 3.4 gives the known manufacturers of coating pans.

The coating drum rotates and the product tumbles in the drum, always remaining near the bottom of the pan, as shown in Figure 3.11. Coating pans manufactured by Vector Corporation and Freund

Figure 3.9. Partially perforated coating drum. Courtesy of Vector Corporation.

Industrial Co., Ltd., are partially perforated, with ductwork built into the outside of the pan. Process air enters the coating pan through a mouth ring at the front opening of the pan, or through the back of the pan, and is drawn out through the ducts built into the outside of the cylindrical portion of the drum, see Figure 3.9. The perforations

Figure 3.10. Fully perforated coating drum. (Courtesy of Thomas Engineering Inc.

Table 3.4. Coating Pan Types and Manufacturers

Manufacturer	Type of Coating Drum	Manufacturing Location	Address
Driam USA	partially perforated	West Germany	Spartanburg, South Carolina
Freund Industrial Co., Ltd.	partially perforated	Japan	Represented by Vector Corp., Marion, Iowa
Glatt Air Techniques	fully perforated	West Germany	Ramsey, New Jersey
O'Hara Technologies Inc.	fully perforated	Ontario, Canada	Ontario, Canada
Thomas Engineering Inc.	fully perforated	Hoffman Estates, Illinois	Hoffman Estates, Illinois
Vector Corporation	partially perforated	Marion, Iowa	Marion, Iowa

Figure 3.11. Product coating. Courtesy of Vector Corporation.

in the coating pan are located over these ducts. This arrangement forces the process air to travel through the tablet bed. Because of the duct system and the tablets, there is a considerable pressure drop in the path of the process air.

The Thomas Engineering Inc. coating pan is a fully perforated drum. Inlet and exhaust ductwork are assembled such that there is a seal to the rotating cylindrical surface of the coating drum. The process air exhaust port is located below the tablets, and the inlet port is above the tablets. This arrangement provides less pressure drop for the process air but allows some air to travel around the product, depending on the efficiency of the seals between the ducts and coating drum.

The Glatt coating pan is also fully perforated. The inlet process air enters through a duct above the tablet bed and is exhausted through a duct below the tablet bed. This arrangement allows air to enter the pan, pass through the tablets, and exit through the perforated areas of the pan, but there is a marked pressure drop.

The Driam coating pan has an arrangement that will allow reverse process air through the tablets, direct air through the tablets (similar to the previously described pans), or reverse process air with the exhaust air being drawn out through the horizontal shaft about which the pan rotates. The O'Hara coating pan is similar to the Thomas Engineering coating pan with respect to the path of the process air.

Coating can be performed with any of these pans if the accessory equipment (blowers, heaters, filters, etc.) is configured properly for the physical layout, and the individual pieces are correctly specified. Some coating pan equipment includes auto discharge (see Figure 3.1, the Driam coating pan). The mixing baffles are configured so that when the pan is rotated in a reverse direction, the product will be discharged out the door. Thomas Engineering Inc. also offers this feature. Many of the other coating pans have a trap door on the product drum so that the product can be discharged into product trays placed under the units. Several manufacturers will supply product loading chutes that will allow the product to be loaded from above the coating pan instead of through the drum opening.

Coating pan capacities for laboratory size units range from 1 L to 110 L. Capacities for production size coating pans range from 90 L to 850 L. The laboratory size units are used to perform feasibility studies and develop product and processes. An example of a laboratory size unit is shown in Figure 3.12. In general, each manufacturer makes sizes of coating pans that are comparable. Table 3.5 shows the size and capacity of coating pans from different

Figure 3.12. Laboratory size coating pan. Courtesy of Vector Corporation.

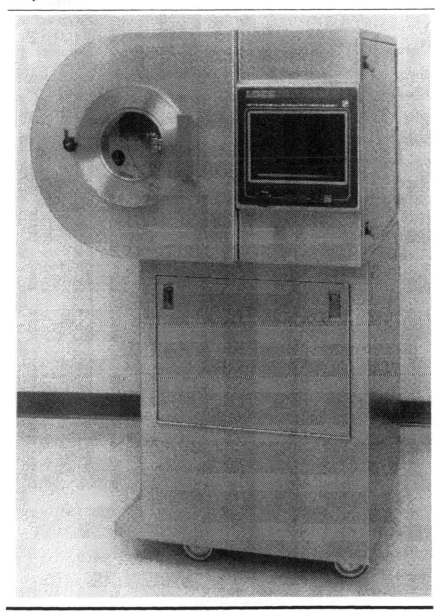

Table 3.5. Production Coating Pan Models and Specifications

Manufacturer	Model	Drum Diameter (inches)	Brim Volume* (liters)	Pan Dimensions (W × D × H; inches)
Driam USA, Inc.	500/600	24	38	63 × 39 × 65
Driam USA, Inc.	900	35	125	47 × 67 × 71
Driam USA, Inc.	1200	47	150	63 × 71 × 75
Driam USA, Inc.	1600	63	500	79 × 96 × 87
Freund Industrial Co., Ltd.	HCF-100	39	90	55 × 63 × 69
Freund Industrial Co., Ltd.	HCF-130	52	225	65 × 68 × 80
Freund Industrial Co., Ltd.	HCF-150	59	350	73 × 78 × 89
Freund Industrial Co., Ltd.	HCF-170	67	550	81 × 88 × 99
Glatt Air Techniques Inc.	750	30	55	47 × 63 × 71
Glatt Air Techniques Inc.	1000	40	110	55 × 71 × 74
Glatt Air Techniques Inc.	1250	50	220	71 × 79 × 79
Glatt Air Techniques Inc.	1500	60	440	75 × 93 × 93
O'Hara Technologies Inc.	FC-36	36	105	48 × 52 × 76
O'Hara Technologies Inc.	FC-48	48	200	64 × 55 × 75
O'Hara Technologies Inc.	FC-60	60	480	76 × 65 × 90
O'Hara Technologies Inc.	FC-66	66	850	80 × 74 × 92
Thomas Engineering Inc.	48S	48	200	60 × 47 × 75

Continued on next page.

Continued from previous page.

Manufacturer	Model	Drum Diameter (inches)	Brim Volume* (liters)	Pan Dimensions (W × D × H —inches)
Thomas Engineering Inc.	60S	60	490	75 × 59 × 81
Thomas Engineering Inc.	66S	66	850	83 × 74 × 86
Vector Corporation	HC-100	39	90	55 × 63 × 69
Vector Corporation	HC-130	52	225	65 × 68 × 80
Vector Corporation	HC-150	59	350	73 × 78 × 89
Vector Corporation	HC-170	67	550	81 × 88 × 99

* Brim volume is defined as the capacity for product up to the opening of the drum.

manufacturers. Each manufacturer will provide specifications for the general operation of their coating pans. Typically, they will recommend and/or furnish the peripheral equipment required to complete the system.

Comparison of Continuous Coating Equipment. Two examples of continuous coating units are shown in Figure 3.13 and 3.14. The product is moved through a drum at a rate that allows the tablets to be completely coated as the product traverses the length of the coating drum. The area where the coating occurs is in a drum that rotates. There are mixing baffles similar to those in the batch coating pans that move the product from one end of the coating unit to the other. These baffles are shown in Figure 3.15. As the coating unit drum rotates, the product tumbles, thus allowing all sides of the product to be coated under the spray. Table 3.6 lists two manufacturers of continuous coating equipment with specifications.

Figure 3.13. Continuous coating equipment. Courtesy of Thomas Engineering Inc.

Air Equipment

All coating pan equipment, including continuous coating equipment, requires conditioned process air to perform the coating. The process air generally should be dehumidified, cleaned, and heated. An air handler is used to perform these functions. Figure 3.16 is a functional diagram of an air handler.

The first section is generally composed of two filters: a sticks and stones or bird type gross filter followed by a prefilter. The prefilter is typically rated for 30 percent efficiency (ASHRAE 1996). The next section is a preheat unit followed by a dehumidification unit. The preheat section is placed before the dehumidification unit to protect any chilled water coil (or refrigeration coil) from freezing. Three methods of dehumidification are used: chilled water, refrigeration, or a desiccant dryer. The desiccant dryer is generally used to obtain lower dew points. If humidification is required (during the winter when the air is very dry), this is accomplished by injecting steam into the process airstream. A heating coil is required to heat the process air.

Figure 3.14. Continuous coating equipment. Courtesy of Vector Corporation.

The air handler shown in Figure 3.16 is a face and bypass unit that allows the unit to switch between cool air and heated air rapidly. A nonbypass type air handler will require some time to cool down, because the heater coils remain in the process airstream after the heat has been turned off. After spraying is completed, the product is normally cooled before being removed from the coating pan. The face and bypass unit can achieve a shorter production time because the heating coil does not have to cool down before the product is cooled. The air handler will contain a blower that must be rated for the proper quantity of process air (cubic feet per minute, CFM) to provide heat transfer into the product. The final stage of an air handler should be a high efficiency particulate air (HEPA) filter

Figure 3.15. Continuous coating drum. Courtesy of Thomas Engineering Inc.

Table 3.6. Manufacturers and Specifications for Continuous Coating Equipment

Manufacturer	Model	Drum Diameter (inches)	Product Throughput (L/h)	Pan Dimensions (W × D × H) (inches)
Coating Machinery Systems, Inc. Huxley, IA (Vector Corp.)	Workhorse	15	up to 1,800	72 × 144 × 97
Thomas Engineering Inc., Hoffman Estates, IL	Continuous Coater	24	up to 1,000	50 × 175 × 118

Figure 3.16. Coating pan air handler. Courtesy of Vector Corporation.

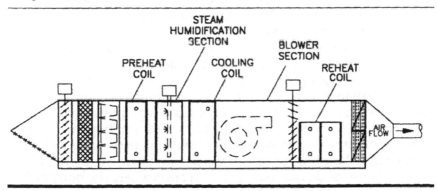

to provide clean air to the coating drum with 99.97 percent efficiency (Flanders Filters 1984).

Figure 3.6 showed the components that may be used in a typical batch coating pan system. The ductwork between the HEPA filter and the coating pan should be stainless steel to preserve the integrity of the clean process air. When another type of ductwork is used, such as galvanized or painted ductwork, chipping can occur (possibly due to rust) and add particles to the product. The quantity of process air is generally recommended by the manufacturer for the size of the equipment. It is desirable to use the maximum amount of process air possible to achieve the maximum heat for the product. However, one must use caution when a decision is made to increase the airflow within the system, because the pressure drops increase exponentially, and turbulence can be generated that will affect the spray patterns inside the coating pan. The inlet blower should be capable of providing sufficient air pressure at the desired quantity of air to bring the pressure in the coating pan to atmospheric pressure or slightly negative (up to -4 in. of water column), when the exhaust blower is operating, and product is loaded in the pan. The exhaust blower must provide a vacuum pressure at the required airflow that will overcome the losses created by the different components between the coating pan and the exhaust blower. This allows the full capability of the exhaust blower to be utilized.

The outlet process air should be ducted to a dust collector to collect solids that did not adhere to the product or dust that is generated when product is loaded into the coating pan. The ductwork before and after the dust collector, and the dust collector itself, must be able to withstand the amount of vacuum that will be created by the exhaust blower. The dust collector also must be designed with an explosion vent that will relieve the pressure in case of a dust explosion. When film coating is performed using a solvent as the carrier, the entire system must be designed to prevent explosions or vent the explosion. Typically, the manufacturer will recommend a system that will satisfy the requirements for a specific type of solvent.

As mentioned earlier, the partially perforated coating pan will create more pressure drop, thus requiring a high pressure blower; a fully perforated coating pan requires less pressure drop but a larger quantity of air. An exhaust blower should be used to provide the recommended airflow for heat transfer and a vacuum of 1–4 in. of water column inside the coating pan. In areas where the noise of the blower cannot be tolerated, especially on systems that require a high pressure blower, a silencer should be used after the exhaust blower. Most manufacturers of coating pans will provide the above peripheral equipment as part of a turnkey system. Some systems, as shown in Figure 3.17, will have a bypass section which allows the process air to bypass the coating pan. This bypass section is used where the coating pan door must be opened for several minutes, thus allowing cool air to enter the coating pan. If the temperature probe that is used for the control loop is in the exhaust ductwork, the control loop will call for a higher temperature. When the coating pan door is closed, the inlet process air temperature will be considerably higher, which can then cause the product temperature to

Figure 3.17. Coating pan system with bypass. Courtesy of O'Hara Manufacturing Ltd.

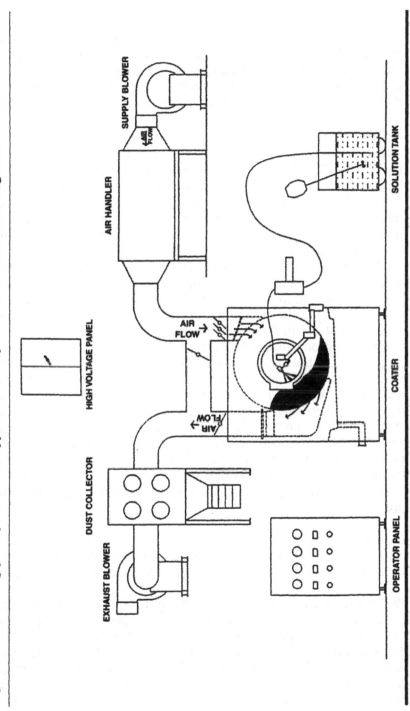

exceed the set limits. The bypass allows the process air temperature to remain constant. The bypass must also have a pressure drop to prevent the airflow from increasing, thus causing a temperature change during bypass conditions. The bypass allows the temperature control loop to hold the process air at a constant temperature during the time the airflow is not passing through the coating pan.

Spray Systems

The design and operation of the spray system is critical for the coating process. In general, two types of spray systems are used: an open loop system or a recirculation loop arrangement, as shown in Figures 3.18a and 3.18b. The open loop arrangement is typically used with a hydraulic spray gun, while the recirculation loop is used with pneumatic spray guns. Figure 3.11 showed the spray guns mounted in a coating pan on a spray gun manifold.

The recirculation loop provides more flexibility, but requires more equipment and tubing to configure. When the spray is off, the coating formulation is recirculated back to the solution tank, providing mixing in the tubing to prevent the precipitation of solids. A separate agitator in the solution tank should be employed for coating formulations prone to precipitation. To use the recirculation loop, valves in the spray gun and the recirculation loop must be provided to change from spray to recirculation. A flow measuring device is not needed for a positive displacement pump, but it is desirable. A separate pump can be provided for each spray gun, or a single pump connected through a manifold can feed all spray guns. The manifold can create different pressures at each gun due to the pressure drops in the manifold tubing, which in turn will cause different droplet sizes and distributions at each gun. Many spray systems use the multiple pump concept to prevent the pressure drop problem.

Comparison of Different Spray Guns. Two types of spray guns are typically used for coating: pneumatic and hydraulic. A third type is an ultrasonic, which has had limited success for coating processes.

Pneumatic Spray Gun. The pneumatic gun sprays liquid through an orifice that is subjected to a stream of compressed air that breaks the spray into droplets. The size of the droplets, the distribution of droplets, and the type of pattern is determined by the quantity of compressed airflow and the physical configuration of the spray gun. Figure 3.19 illustrates a pneumatic spray gun and shows the different ports for the air and solution. Manufacturers of spray guns are

Figure 3.18a. Recirculation (closed) loop spray system. Courtesy of Vector Corporation.

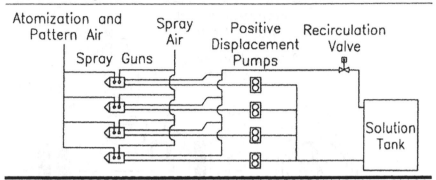

Figure 3.18b. Open loop spray system. Courtesy of Vector Corporation.

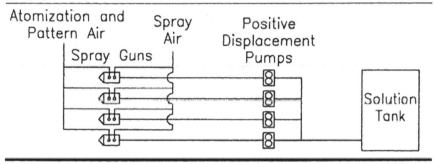

shown in Table 3.7. Solution and air orifice sizes on the spray guns may have to be changed for different types of coating suspensions; in general, however, a single orifice size will be sufficient for the typical coating suspensions that are used.

Hydraulic Spray Gun. A hydraulic spray gun uses pressure on the suspension to force it through an orifice that shears the liquid into droplets. The orifice must be changed for different types of spray patterns and spray distributions. Hydraulic spray guns have limited use in coating pans, because the spray distribution is considerably wider than with pneumatic spray guns.

Comparison of Spray Guns. In both types of spray guns, as the spray rate is changed, the pattern, droplet size, and distribution will vary. The pneumatic gun provides more flexibility because the air can be

Figure 3.19. Pneumatic spray gun. Courtesy of Vector Corporation.

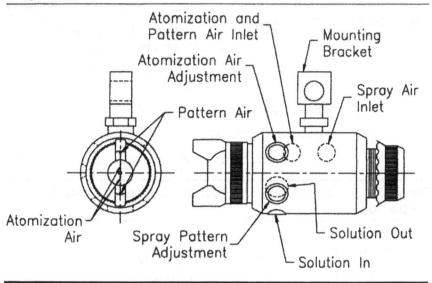

Table 3.7. Spray Gun Manufacturers

Manufacturer	Address	Trade Name	Types of Spray Guns
Binks Sames Corporation	Geneseo, IL 61254	Binks Guns	Pneumatic
Freund (Vector Corp.)	Marion, IA 52302	Vector or Freund Guns	Pneumatic
Schlick	Coburg, Germany 96540	Schlick Guns	Pneumatic Hydraulic
Spraying Systems Co.	Wheaton, IL 60189	Spraying Systems	Pneumatic Hydraulic

varied to obtain the desired spray without changing the nozzle. The ultrasonic spray gun, on the other hand, provides a small droplet size over the range of spray rate. Typically, with the coating suspensions currently being used, the small droplet size causes difficulty in the coating process due to spray drying of the droplets and turbulence in the coating pan. The big advantage of the ultrasonic spray gun is that a large orifice is used, which results in less nozzle plugging. Both the hydraulic and pneumatic spray guns are subject to plugging, causing disruption of the droplet size distribution and the spray pattern. The hydraulic gun is prone to more plugging than the pneumatic spray gun. The pneumatic gun will not drip after the solution is turned off if the air is left on for a short period of time.

Suspension Delivery Pumps

The open loop arrangement described earlier for the spray system is generally used with a positive displacement pump, such as a gear pump, a lobe pump, or a peristaltic pump (Figure 3.20). These pumps are capable of supplying pressures from 15 to 100 psig to the spray nozzles. Peristaltic pumps have been used on a limited basis because of the high flow rates that are needed for pan coating (except in laboratory coating pans). Centrifugal pumps are normally not used due to pressure requirements and the need for a flow measurement device.

Pump selection is determined by the type of spray gun to be used. Hydraulic spray guns require 30–60 psig of solution pressure at the desired flow rates. Pneumatic spray guns require 1–15 psig at the desired flow rate. Ultrasonic spray guns require only enough pressure to supply the solution to the spray head, which is usually less than 10 psig. Normally, pumps that operate at lower pressures create fewer problems relative to pump wear and sealing of the connections to the pump and spray guns. To ensure that proper flow and pressure are provided to each spray gun, many of the spray systems incorporate a separate pump for each spray gun. A single pump or a single tubing distribution system to the spray guns will result in different pressures and flow rates at the spray guns. When selecting a pump, one must consider the solution source location. Some coating formulations may require additional pump pressure due to a high vacuum pressure on the inlet side of the pump. When using more than one pump, the inlet tubing must be sized for the total number of pumps and the vacuum loss due to the size of the inlet tubing.

The connections for the spray system are very critical and must be installed with care. If there is a leak in the tubing on the inlet side

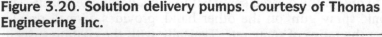

Figure 3.20. Solution delivery pumps. Courtesy of Thomas Engineering Inc.

of the pump, air bubbles are created in the suspension and travel to the spray heads, causing distortion of the spray pattern and droplet size distribution. Because all spray guns are mounted inside the coating pan over the product, any leaks in the tubing, piping, or connections will result in dripping onto the product, causing twinning (tablets stuck together) or peeling (tablets stuck together and separated).

Delivery Control. The coating on the tablets can only be consistent if the amount of suspension being sprayed is consistent. For this reason, the flow of suspension should be controlled. In the past, assumptions were made that if a positive displacement pump is used, the rotational speed of the pump shaft (rpm) can be used to determine the flow rate. However, because of pressure changes in the system and where more than one pump is used, variations in the spray gun outputs can occur. Therefore, it is advisable that some form of measurement and control be used for the suspension flow.

The most common form of measurement is to use a mass flowmeter. Spray systems that use a manifold for the spray guns can be configured to use a single pump, but a mass flowmeter should be used at the outlet of the pump. Spray systems that use multiple pumps feeding individual guns should incorporate a mass flowmeter in the common source line feeding the pumps. There are several types of mass flowmeters, including vortex flowmeters, coriolis mass flowmeters, and turbines. Turbines are typically not used because they cannot be cleaned properly without removing them from the coating suspension line. Differential pressure is not used because this requires an orifice plate that generates more pressure drop for the spray system, thus increasing the workload for the pump. These flow devices will normally have to measure mass flow over a range of 100 to 1,000 cm^3/min.

The second best method to measure the application of the coating is to use a weight loss system. Such a system involves putting the solution tank on load cells (weight-measuring sensors) and then, knowing the density of the suspension, the amount of material applied can be determined as a reduction in the tank weight. This weighing system must be capable of setting the tare weight at zero before spraying is started.

System Controls

The system controls may consist of one of several options and should be chosen based on the tolerance required for the product, the operator's technical knowledge level, and the amount of automation desired. There are basically three types of systems:

1. Manual: All functions are selected and controlled by an operator.

2. Semiautomatic: These systems use a combination of manual and automatic systems.

3. Automatic: This system will require a minimal amount of operator input during the coating process.

Typically, a process is developed such that the equipment must be operated at specific conditions. The following conditions should be held constant:

- Process airflow.

- Process air temperature into the coating pan.

- Average static pressure in the coating pan.

- Humidity (or dew point) in the process air going into the coating pan.

- Coating pan rotation (rpm).

- Spray rate of the suspension.

In most systems today, the process airflow is controlled by a variable frequency drive for the inlet blower motor and the exhaust blower. Until the variable frequency drives became cost-effective, the process airflow was adjusted using dampers in the ductwork, and the blowers were run at full speed. The dampers can cause problems for the coating process by introducing turbulence in the air patterns in the coating pans. Variable frequency drives can be configured to operate as a local control or as a remote control unit. The airflow is typically measured in the inlet duct, because the air is clean at this location, thus requiring minimum sensor maintenance. The airflow can be measured using a pitot tube type flowmeter (measures air pressure converted to flow), a vortex flowmeter, a turbine flowmeter, or other air-measuring devices that can be located in the ductwork. Ideally, the measuring device should compensate for temperature changes of the process air. Measurement of airflow in the inlet duct assumes that the coating pan is closed during the coating process and that there is minimal leakage into or out of the coating pan or exhaust ductwork.

An analog output must be supplied from the flow-measuring device for a programmable logic controller (PLC) to control the airflow. A variable frequency drive connected to the inlet blower is used in the semiautomatic or automatic modes to provide constant airflow and pressure. A tube located in the coating pan measures the static pressure. The exhaust air blower speed is controlled by the variable frequency drive from a controller that is connected to the static pressure. This type of control will occur for the semiautomatic and automatic modes of operation. In manual mode operation, variable frequency drives are adjusted by an operator at a local or remote terminal.

The process air temperature is generally measured at both the inlet and at the exhaust of the coating pan with a resistance temperature detector (RTD) probe or a thermocouple. The temperature can be held constant with a temperature controller that controls a steam valve or electric heater in the air handler unit. If all of the other parameters listed previously are held constant, the exhaust temperature will eventually stabilize at a constant temperature. The exhaust temperature will be approximately equal to the product temperature, depending on heat losses between the product temperature and the measurement probe. The exhaust temperature response will be approximately the same as the product temperature. Ideally, a constant product temperature is the preferred method for control; however, this temperature is difficult to measure because of the motion of the product. Some systems have a controller, when used in the semiautomatic or automatic modes, that operates in a cascade manner, such that the inlet temperature setpoint is the controlling parameter until the inlet temperature is reached, at which time it switches to the exhaust temperature for control. The temperature controller is also required in the manual mode of operation.

The amount of moisture in the system should be measured in the inlet process stream but after the chilled water coil (or refrigeration coil). A dew point sensor should be used at this location, as a relative humidity probe will show 100 percent because the air is generally cooled to below the saturation point. The dew point measurement should then be used to control the dehumidifier or humidifier. Except for automatic systems, the dehumidifier is operated at full dehumidification in the summer and could be turned off in the winter when the outside air is naturally dry. This also prevents freezing of chilled water coils (if used) and reduces utility costs. The coating process must be developed so that small changes in humidity will not affect the coating.

In most coating pans, the drive motor for pan rotation is controlled by a variable frequency drive. In the manual mode, the speed of the pan (rpm) is adjusted by local or remote operator panels. In the semiautomatic and automatic modes, an instrument will measure the rpm of the pan and use a controller to adjust the rpm to a setpoint and hold it constant.

As discussed earlier, a controller connected as a mass flowmeter should be used to provide a constant flow rate of the suspension to the spray guns. In a manual system, the rpm of the pump can be measured and adjusted by the operator for a constant flow rate.

Many other parameters can be measured for troubleshooting purposes. Differential pressures can be measured across inlet filters to signal when filters become dirty. In the closed loop spray system, the back pressure on each spray gun can be measured to signal a plugged gun. The air for pneumatic spray guns should be measured or controlled to ensure proper atomization of the solution. Compressed air sources for valves, atomization, and other instruments can be monitored.

A semiautomatic system will generally use a separate controller for each parameter measured, which should be held constant. An operator will adjust the setpoints for each controllers, which may have to be changed during the coating process.

The automatic system allows the operator to input a recipe that will be used to control the coating process. An operator has to start, stop, load, and unload the product. The setpoints are adjusted based on the stored recipe. The operator will be signaled if any of the troubleshooting parameters indicate an erroneous condition. The automatic system will contain a PLC or a computer processing unit (CPU) that is programmed to accept measurements, make decisions, and control the peripheral devices. The human/machine interface can be analog and digital meters or a video display with a full or dedicated keyboard. Analog signals are provided by the instruments or sent to the controlling devices (valves, setpoints, etc.). Digital signals are provided by instruments that have a fixed setpoint and are used to signal erroneous conditions or for starting and stopping the equipment.

All three control systems require the measurement of certain parameters in order to make any coating system operate with consistent results. The manual system has instruments that provide only local readouts, which are used by the operator to adjust the controls. Both types of automation require outputs that can be input into a PLC or computer.

The following measurements are the minimum required for any coating system in order to provide a consistent coating operation:

- Inlet process air temperature.

- Exhaust process air temperature from the coating pan.

- Process airflow volume or flow rate in the inlet duct to the coating pan.

- Pump speed or suspension spray rate.

- Coating pan rotation (rpm).

The following controls are required for each system:

- Power to the system.

- Temperature controller.

- Airflow control (manual or motorized dampers or blower motor speed control).

- Solution volume or rate control (mechanical or pump motor speed control).

- Coating pan rotational speed (mechanical or pan motor speed control).

Both measurements and controls will be analog in nature, either with an analog meter or an analog signal to a PLC or computer.

Other measurements that may be included in a coating pan system are as follows:

- Process air temperature after the preheater.
- Dew point temperature after dehumidification or humidification.
- Differential pressure across the inlet filter to the air handler.
- Differential pressure across the HEPA filters in the inlet air duct.
- Differential pressure across the coating pan.
- Static pressure in the coating pan.
- Differential pressure across the dust collector.
- Static pressure after the air handler.
- Static pressure at the inlet of the exhaust blower.
- Product temperature.
- Main compressed air pressure.
- Solution tank weighing system.
- Solution pressure.
- Solution mass flowmeter.
- Compressed airflow rate to pneumatic spray guns.

Analog Inputs/Outputs. All of the controlling parameters used to provide constant conditions must have an associated analog input and output. Analog output is required for each controlling device (e.g., dampers, steam valves, variable frequency drives, and pneumatic valves). The following items are other parameters that may have analog inputs and outputs:

- Process air temperature after the preheater.
- Dew point temperature after dehumidification or humidification.
- Differential pressure across the coating pan.
- Static pressure in the coating pan.
- Product temperature.
- Solution tank weighing system.

- Solution mass flowmeter.

- Compressed airflow rate to pneumatic spray guns.

The inlet blower speed or the inlet damper is controlled based on the process airflow. The exhaust blower or damper is controlled based on the static pressure in the coating pan. The humidifier or dehumidifier is controlled based on the dew point values. The steam or electrically heated preheat coils are controlled based on the process air temperature after the preheat coil. Either the tank weighing system or the solution mass flowmeter is used to determine when the proper amount of suspension has been applied to the product as well as the application rate. The application rate is used to control the pump for the suspension. The main steam heater valve or electrical heater is controlled by the inlet air temperature. Some systems use the exhaust to control the temperature of the process air. The best control for the process air temperature is to use a cascade control, whereby heating is controlled by both the inlet process air temperature and the exhaust temperature. The product temperature can be substituted for the exhaust process air temperature (if it is available). The exhaust process air temperature (or product temperature) is not usually used as the only temperature control parameter because of the slow response caused by the heat sink of the product and coating pan. This situation will either create a long preheat time, if the temperature controller is tuned properly, or cause a large overshoot of the inlet process air temperature when the controller is tuned to reduce the preheat time. The compressed airflow rate to each gun is measured when pneumatic spray guns are used. The flow rate can be adjusted manually with a needle valve on the spray gun.

Digital Inputs/Outputs. The remainder of the measured parameters are used as digital inputs/outputs to sense when an erroneous condition has occurred in the process parameters. The following items will have digital inputs and outputs for alarms and mode control:

- Differential pressure across the inlet filter to the air handler.

- Differential pressure across HEPA filters.

- Differential pressure across the dust collector.

- Static pressure after the air handler.

- Static pressure at the inlet of the exhaust blower.

- Main compressed air pressure.

- Suspension pressure.

These parameters are used to alert the operator or place the coating system in a neutral state. The neutral state is normally defined as the condition with the spray off, the temperature set to a reduced temperature, and the process airflow continuing at the normal rate. When an alarm occurs, and the system is placed in a neutral state, the operator must assess the situation and correct the situation or stop the coating operation.

Differential Pressure Gauges. Differential pressure gauges are used to monitor functions, such as dirty filters, proper differential pressure across the tablet bed (or static pressure on the exhaust side of the coating pan), and proper operation of the dust collector. Some differential pressure gauges measure static pressure, which is the differential pressure between the function being measured and atmospheric pressure. These are local gauges or instruments providing electrical output for maintenance purposes.

Airflow Meter/Station. The airflow meter/station used to measure the inlet airflow measures the rate in feet per minute. Because the airflow is in a duct with a known cross-section at the point of measurement, the reading can be converted to mass flow in standard cubic feet per minute (SCFM)—the flow rate in feet per minute times the cross-sectional area of the duct in square feet. This measurement is dependent on the temperature of the air, unless the instrument is corrected for temperature. If the instrument corrects for air temperature, then actual cubic feet per minute (ACFM) is measured. Instruments that have been successfully used to measure airflow on coating systems are pitot tubes (not compensated for temperature) and vortex flowmeters (independent of temperature changes). The most popular airflow-measuring instruments are airflow stations with a built-in flow straightener. This station is composed of a series of pitot tubes that provide the average airflow of air distribution inside the duct. These instruments should be located in the ductwork, with a length of at least five diameters of straight duct before and after the instrument. This generally helps to provide a symmetrical distribution of the airflow in the duct.

THE COATING PROCESS

General Description

Approximately 40 to 60 percent of water is evaporated between the spray gun and the tablet bed when the coating solution is sprayed on the product surface. The remaining water evaporates from the tablet surface because of conduction of heat from the hot tablet surface. This is approximately 10 times faster than the conduction of heat from air to the coating solution droplet. Hence, the viscosity of the coating solution droplets changes if and when the distance between the spray gun and the tablet bed is changed. An acceptable film coating will form only upon coalescence of droplets if the coating solution droplets have the proper viscosity.

Typical Coating Process Steps

If this is the first time that a particular product has been coated, it is a good idea to estimate the approximate batch size to use. Coating pans are usually rated for a specific brim volume capacity. Brim volume is the volume capacity of the pan if the pan is loaded to the very bottom of the pan opening. The initial pan load is usually somewhat less than the brim volume, which prevents product from spilling out of the pan opening in case the product volume increases or if the product movement should change during the coating process. As a starting point, the brim volume can be multiplied by 95 percent to give the pan load volume, which can then be multiplied by the product's bulk density (in g/cm^3) to determine the approximate maximum batch weight. The product's bulk density can be approximated by determining the weight of 1 L of tablets. Batches of 70–75 percent of the maximum can usually be coated without a problem. For smaller batch sizes, it is necessary to examine the product movement to verify that it is adequate. It may also be necessary to reduce the exhaust opening to prevent air from being drawn around the tablet bed rather than through it.

For continuous coating pan systems, the throughput is controlled by a combination of the pan angle and the pan speed (rpm). A higher throughput is achieved by increasing the pan angle or the pan speed. Either of these changes causes the product to flow more rapidly through the pan, thus increasing the potential production rate. One continuous coating pan system allows the drum to be adjusted from a -2° to 6°.

The typical coating process consists of several different steps. The first step is to verify the proper operation of the spray delivery system. The spray system must be calibrated to deliver a consistent gun-to-gun delivery of the coating solution. This is especially critical for systems that have solution lines that are manifolded together. With this type of arrangement, a restriction or difference in pressure drop between the spray guns will result in a nonuniform solution delivery. Calibration can be accomplished by individually collecting and weighing the solution from each spray gun for a set interval of time. The calibration procedure should determine if the overall delivery of solution is accurate (if it matches the theoretical rate) and the variation in solution flow between guns. If there is an unacceptable variation (usually $> \pm 5.0$ percent), then the system needs to be adjusted so that the pressure drop between the guns is the same.

Most spray guns are equipped with a needle valve adjustment cap that controls the clearance between the spray needle and the liquid nozzle. If a spray gun delivers more than its prescribed quantity of solution, the adjustment cap should be turned clockwise to restrict the flow to that particular gun. After making an adjustment to the spray guns, the solution collection should be repeated. When uniform solution delivery has been achieved, it is necessary to calibrate the nozzle air volume to the spray guns. Some spray guns have a common line for providing atomization air and pattern air. For these systems, the air volume is achieved by setting the desired nozzle air pressure. Other guns have separate controls for the atomization and pattern air, and a mist checker or flowmeter for displaying the actual air volume. For these systems, a needle valve (located on the spray gun body) is usually adjusted to set the atomization air volume. Once the atomization air volume has been set for all of the guns, a second needle valve is adjusted to the desired volume for pattern air. After these air volumes have been set, it is possible to verify the approximate dimensions of the spray pattern by passing a hand through the airstream. This spray pattern should be checked at a distance that is equal to the distance between the tablet bed and the spray guns.

Loading/Charging

After proper setup of the pan has been confirmed, the pan can be loaded with product. Product can be loaded into the pan either through the front of the pan (pan mouthring) or through a discharge door located on the flat of the pan (if it is so equipped). During pan loading, the exhaust air is usually turned on to eliminate or minimize exposure to irritating and potentially hazardous dusts. It is a good idea to minimize the distance that the product is allowed to drop during charging, as this will minimize or eliminate tablet breakage. It is usually recommended that the pan be jogged occasionally during loading to move the product toward the back of the pan.

Preheat/Dedusting

After loading, preheat the product to the desired process temperature. While the product is being preheated, it is also being dedusted by the process airflow. This is also a good time to circulate coating solution through the spray guns (if a recirculation system is used). Verification that the spray guns are positioned correctly can be accomplished by checking both the angle to and the distance from the tablet bed. During the preheat step, the product should be rotated continuously at a very slow speed or jogged intermittently. This will minimize attrition while ensuring that the product is uniformly heated. For products that are heat sensitive, it will be necessary to jog on a more frequent basis to prevent overheating of the upper surface of the tablet bed. Tablets are generally preheated to a specific product or exhaust temperature. The product and exhaust temperatures are not ordinarily the same, although during the coating process, they tend to be fairly close. Once the exhaust air or product has been heated to the desired starting temperature, coating can be initiated.

Seal/Barrier Coat

In a few rare cases, it may be necessary to apply a barrier coating prior to the application of the film coating. An example of this is when there is an interaction between the film polymer and the product substrate (e.g., interaction between an enteric polymer with phthalate groups and tablets containing an alkaline drug [FMC Corp. 1986]). This problem can be eliminated by applying a seal coat of an inert film polymer, such as HPMC prior to the enteric polymer coating. A second example is when there is an interaction between the coating solvent and the product (e.g., an aqueous coating and an effervescent product). To remedy this problem, a seal coating with an alternative solvent may be used. It is possible to coat some effervescent products with aqueous solutions if the spray rate for the initial phase of the coating is reduced and the process temperatures are substantially increased.

Film Coating

To begin the spray cycle, the pan is first set at the desired rotational speed. If a variable frequency drive is used, it is important to allow approximately 5 sec for the pan to achieve the rpm setpoint. Once the product has reached the required speed, spraying can be started. Initiation of spraying will cause the product and exhaust temperatures to drop slightly due to evaporative cooling. If these temperatures drop below the desired range, the inlet temperature must be increased. If the coating system has an inlet air handler with a humidity controller, there will be consistent batch-to-batch correlation between the inlet temperature necessary to achieve the desired exhaust and/or product temperatures. Some systems automatically control the inlet to maintain the needed exhaust temperature. The time required for the coating cycle is determined by a number of factors, including desired coating weight gain, coating efficiency, coating solids level, spray rate, and the size of the spray zone. Since these factors tend to be constant for a particular product, the end of the spray cycle is usually controlled by either time or the application of a set quantity of coating suspension. If the spray cycle is controlled by suspension quantity, this can be accomplished with either a mass flowmeter or a mass balance. Either of these methods will indicate the amount of suspension applied.

The film coating process in a continuous coating pan system is somewhat different than that for batch systems. While the exact process may vary from one pan manufacturer to another, the operational sequence given in Table 3.8 incorporates many of the steps commonly used with these systems.

Gloss Coat

After the primary coating has been completed, a dilute overcoating may be applied to prevent the tablets from blocking or sticking (such as with some aqueous dispersions that are thermoplastic) or to provide a higher film gloss. If the purpose is to provide a higher gloss, a dilute HPMC solution may be used. The process temperatures may be reduced slightly to reduce the amount of spray drying that occurs.

For some aqueous coating dispersions, it is recommended that the product be maintained at a slightly elevated temperature to cure the coating fully and to provide a stable release profile. After coating, the product should be cooled prior to the application of a powdered wax. Cooling the product after coating also minimizes potential instability problems due to heat sensitivity of the active. The product temperature will rise after the spray is turned off due to loss of the evaporative cooling effect. Product is typically cooled to an exhaust or product temperature of 25–30°C.

Wax Addition

After the product is cooled, a powdered wax may be used to provide a higher tablet gloss. The waxes typically used are either a carnauba wax or a combination of carnauba and beeswax. A small quantity of wax (5–10 g/100 kg of tablets) is applied to the rotating tablet bed.

Table 3.8. Film Coating Process—Continuous Coating Pan System

Step	Procedure
1. Set the Drum Angle	The drum angle is set based on the product shape, product density, and the desired production rate.
2. Preheat the Product	Turn on the inlet blower, exhaust blower, and air heater. The air heater is set to the desired temperature setpoint.
3. Set the Pan Speed	Pan speed is set based on the product shape and desired throughput.
4. Set the Pump Rate	The pump system is set to deliver the desired spray rate.
5. Start the Product Flow	A continuous coating system incorporates the use of a product feed device, such as a vibratory feeder, to deliver tablets to the drum at the desired rate.
6. Start the Spray	The spray cycle is started once the product mass reaches the desired temperature. The liquid flow will be staged to the spray guns. The spray is initiated one gun at a time, beginning with the gun nearest to the product inlet. Spray does not begin until product is present in that section of the coating drum.
7. Stop the Spray	Once the coating is completed, the product feed system is shut off. The guns are shut off in an order reverse of the startup. This minimizes product loss during the shutdown process.

The tablets are allowed to rotate for approximately 5 min with no process air. Then the tablets are rotated for an additional 5 min with the process air on. This allows any excess wax to be exhausted. A canvas-lined pan is not necessary for the wax application, although it may provide a slightly higher gloss.

Product Discharge

At this point, the product is ready for discharge. Depending on the type of pan used, the product will be discharged through a "trapdoor" on the flat of the pan or through the front pan door. If product is discharged through the front door, a scoop may be temporarily installed to convey product through the front door. In other systems, the pan may be rotated backward or counterclockwise to discharge the product, in which case the baffles are designed to direct the product out the front door. Discharging of the entire batch is usually achieved in 2–10 min, depending on the pan size and type of discharge method employed. For most cases, the coated product should not be allowed to drop more than 2–3 ft, otherwise the tablets may break.

Process Parameters/Coating Dynamics

Process Airflow

Volume. In all coating pans, elevated temperatures are used in conjunction with the airflow to convert the solvent in the coating suspension into a vapor and drive it away from the tablets being coated. In general, the amount of water vapor that can be removed is directly proportional to the air volume that passes through the coating pan and is limited by the saturation of the airstream. Therefore, if maximum evaporative capacity is the rate limiting factor with respect to the spray rate, it would be advantageous to use as high an airflow as possible without creating spray turbulence or air leakage. Manufacturers of the coating pans will usually specify the maximum airflow that can be used and still maintain an acceptable level of turbulence. As the process airstream is heated, its capacity to hold water vapor increases. Again, if maximum evaporative capacity is the rate limiting factor, the inlet air temperature should be as high as possible.

However, spray rates (per spray gun) are frequently limited not by evaporative capacities but by the diminishing quality of the spray as the spray rate is increased. The total spray rate is usually limited by the number of spray guns and, more accurately, by the size of the spray zone available. Therefore, the airflow capacity between pans of different sizes should be in direct proportion to the spray zone. For example, if a coating process is developed in a pan with 2 spray guns (each gun possessing an 8 in. spray pattern) and a process airflow of 600 CFM, scale up to a larger pan with 4 spray guns should have an airflow of approximately 1,200 CFM. Using an airflow in direct proportion to the spray rate will allow a closer correlation between the process temperatures in the different size pans.

Humidity. The process air used in a coating pan is either conditioned or unconditioned. In either case, there are day-to-day variations in the moisture present in the airstream. Humidity in the inlet airstream is usually determined by the dew point temperature, which provides a direct indication of humidity in the air. Relative humidity can be used; however, the temperature of the air must also be known to determine the level of moisture. In the case of conditioned air, variations are less. With unconditioned air, these variations are large and can cause coating problems. If the air is unconditioned, the spray rate must be selected such that if the most humid condition occurs, the film coating would still dry at a rate that would not result in overwetting or stability problems with the product. Since most batches are coated at lower humidities, some level of spray drying would occur. The severity of the spray drying would be directly related to variation in the ambient humidity. Therefore, the growing trend is to dehumidify the process airstream.

Dehumidification not only provides the coating process with more consistent coating conditions, it also provides for greater evaporative capacity. This allows the system to evaporate more moisture at a given airflow and temperature than an airflow of a higher humidity. With the advent of sustained-release coating, some companies have opted for a combination of dehumidification/humidification that allows the process to be conducted at a consistent inlet dew point temperature regardless of the ambient conditions. A system that conditions the inlet airstream to a consistent dew point allows the coating process to operate at a reproducible drying rate, regardless of fluctuations in ambient conditions.

Process Temperatures

As stated earlier, the coating process can be successfully controlled by either the inlet, exhaust, or product temperatures. Process control based on inlet air temperature control is the most common. Control based on exhaust or product temperatures will often react slower because of the heat sink effect of the tablet bed. With inlet temperature control, the exhaust temperature will drop slightly after the spray is started, due to evaporative cooling. This will not occur with exhaust temperature control since the inlet temperature will be controlled to maintain the desired setpoint. In any event, control via any of these methods will yield the same approximate temperatures. Since the moisture from the spray droplets is dried by both convection (due to the process air) and conduction (due to the product temperature), all of these temperatures are equally important. The desired exhaust temperature is dependent on several factors.

Coating Solution Characteristics. The tackier the coating as it dries, the higher the exhaust or product temperatures must be to prevent overwetting defects during the coating process. If the film polymer is thermoplastic, the product must be kept below the temperature at which the polymer begins to soften to prevent the tablets from blocking or sticking together.

Product Temperature Limits. If a product exhibits instability problems at elevated temperatures, the product must be held below these limits. This is especially critical during preheating of the tablet bed, since evaporative cooling of the bed is not occurring. The product may not be uniformly heated if the pan is not being continuously rotated. The product temperature will begin to rise immediately after the spray is stopped due to loss of the evaporative cooling effect. Therefore, it may be necessary to begin a cooldown cycle immediately after the spray cycle.

The inlet temperature required to achieve the desired exhaust temperature will be affected by the spray rate, the percentage of solids in the coating suspension, the heat loss across the pan, and the condition of the drying air.

Pan Speed

To optimize film coating quality, the tablets must be mixed so that each tablet has the same probability of being in the spray for an equal duration of time. It is essential that the product be examined to ensure that mixing is uniform. Mixing problems that occur include sliding of the tablets (usually seen with large capsules or oval-shaped tablets), "dead spots" or sluggish product movement (an extremely serious problem if it occurs while the tablets are in the spray zone), or product thrown into the spray zone by the mixing baffles.

If product movement is not uniform, the first course of action is to evaluate the product flow at different pan speeds. The pan speed selected should be the lowest speed that produces a rapid and continuous product flow through the spray zone. This will allow for the uniform application of a film coating while subjecting the tablets to a minimal amount of abuse. In general, if tablet hardness is a minimum of 6-7 kp (kilopond) (for an explanation of tablet hardness, see "Formulation Requirements: Hardness and Friability") and friability is less than 0.5 percent, tablet attrition will not be a problem. A smaller tablet can be slightly softer, since these tablets produce less abusive tumbling action.

Product flow can be evaluated either subjectively or through mixing studies using tablets of contrasting colors. Tablets of different colors can be placed in different zones (e.g., front, middle, and back) of the pan. Tablet samples are taken at set time intervals to determine the length of time required to achieve a homogeneous mixture. A more sophisticated means of evaluation includes the use of radioactively marked tablets and a counter mounted on the spray bar to record the number of passes through the spray zone per unit time (Leaver et al. 1985). The product is generally assumed to be traveling at the same linear velocity as the inside circumference of the coating pan.

In a continuous coating pan system, product flow is from the charge port to the discharge port. Depending on the coating level, the product may pass through multiple drums (which may be positioned end to end or stacked horizontally). One continuous coating pan system currently available contains 5 spray guns for a 5 ft long coating drum.

Spray Application

Spray Rate. The selection of the proper spray rate is dependent on more than just thermodynamic considerations. If this were not so, the ultimate coating system would offer unlimited airflow and temperature. The spray rate (per spray gun) is also dependent on the spray gun's ability to produce a consistent droplet size distribution. It has been shown that the droplet size distribution will increase as the spray rate is increased. Other factors that must be considered when determining the spray rate are as follows:

- Suspension viscosity: As the viscosity is increased, the spray gun's ability to produce an acceptable droplet size distribution is diminished. Viscosity limits the maximum spray rate that can be used to produce acceptable film quality.

- Spray pattern width: If the spray pattern width is set up properly, then the spray pattern will essentially be the same as the spray gun spacing. The typical gun-to-gun spacing is 5–8 in. At spacings greater than 8 in., the uniformity of the spray across the pattern begins to deteriorate. Spacings of less than 5 in. are an ineffective use of spray guns and tend to add more expense for added spray guns, solution lines, and pumps. There is a limit to how much spray can be applied to a tablet per pass through the coating zone before the tablet begins to exhibit coating defects. Therefore, the wider the spray pattern width (without overlapping adjacent spray patterns), the greater the spray rate per gun.

- Product movement: The more consistent the product flow through the spray zone, the higher the spray rate that can be delivered and still achieve an acceptable level of film coating uniformity. Product movement is often dictated by the pan speed, baffle design, and tablet size and shape.

Droplet Size Distribution. One of the most critical aspects of coating is the manner in which the suspension is applied to the tablet. The spray droplets can be almost any size if the size distribution is sufficiently narrow. If the system is set up to operate so that smaller droplets dry properly, then larger droplets will stay wet and picking or twinning will occur. On the other hand, if the system is set up to dry larger droplets properly, then small droplets will dry quickly and will not spread properly on the tablet, thus causing "orange peeling" and a dull and rough film appearance. Most spray guns used today have a limited range over which the spray can be varied and still maintain a uniform distribution. See Figure 3.21 for examples of some typical droplet size distributions. These droplet size distributions were produced using a film suspension with a viscosity of 130 cps. Ideally, a spray gun that produces a narrow droplet size distribution should be used. This distribution can vary with changes in solution viscosity, spray rate, or a change in the type of solids; therefore, manufacturers of spray guns will generally not provide information concerning these parameters. A simple method of examining this distribution can be performed by quickly passing a sheet of paper through the spray and subjectively analyzing the droplet size. A typical droplet distribution usually ranges from 5 to 250 μm. In general, if droplets vary in size from 30 to 1600 μm, it will be difficult to optimize the film coating quality.

Coating Zone/Pattern

As mentioned earlier, the larger the spray zone per spray gun, the higher the maximum spray rate that can be used per spray gun. Conversely, if the gun spacing or the spray pattern is reduced, then the spray rate should be reduced proportionally. Applying more spray per unit area of the tablet bed beyond a certain point will change the film coating appearance. If the spray rate is increased

Figure 3.21. Droplet size distribution using various spray rates. Courtesy of Vector Corporation.

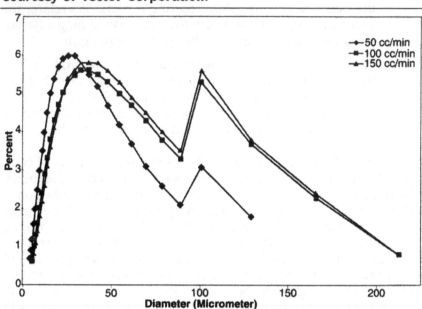

above this maximum, either overwetting or spray drying will occur, depending on whether the process is run dry or wet.

A recent trend in coating is to achieve greater production through using elongated or "stretch" coating pans. Lengthening the coating pan not only increases its capacity but, more importantly, the spray zone. This allows more spray guns to be used, thereby increasing the overall spray rate. An increase in pan volume achieved by increasing the bed depth with no increase in spray zone does not increase the overall spray rate. Therefore, one means of scaling up the process time for coating can be achieved by using the following equation:

$$\text{spray time (large pan)} = \text{spray time (small pan)} \times \frac{\text{batch size (large pan)}}{\text{batch size (small pan)}} \times \frac{\text{spray zone (small pan)}}{\text{spray zone (large pan)}}$$

This equation assumes that the coating zones for each pan are used efficiently.

A number of factors have an affect on the spray pattern width. As the pattern air volume for the spray gun is increased, the spray pattern is widened. An increase in the atomization air volume will cause a reduction in the pattern width. This is logical because the increased volume of atomization air makes it more difficult to flatten or widen the pattern. Lastly, as the spray rate is increased, the spray pattern is widened. Therefore, if the atomization air volume or spray rates are adjusted, reexamine the spray pattern width to ensure that the spray zone is effectively utilized.

Coating Analysis

The following is a theoretical analysis of the film coating process:

Product Specification:

Tablet: 0.32 in. diameter

Weight: 250.0 mg/tablet

Density: 0.73 g/cm^3

Coating Suspension Specifications:

Solids: 12 percent

Density: 1.0 g/cm^3

Quantity: 62.5 kg

Equipment Specifications:

Pan diameter: 67 in.

Pan volume: 550 L

Pan speed: 5 rpm

4 Spray guns

Coating Conditions:

Spray rate: 125 g/min/gun

Total spray rate: 500 g/min

Spray time: 125 min

Assumptions:

Spray zone:

Per spray gun = spray width × spray pattern length = 3 in. × 8 in. = 24 in.^{2}

Total spray zone = number of spray guns × spray zone per gun = 4 × 24 in.^{2} = 96 in.^{2}

Tablets per coating zone = total spray zone × number of tablets/in.^{2}

Assume 22 tablets/in.^{2} in spray zone

96 in^2 × 22 tablets/in.^{2} = 2,112 tablets/coating zone

Total spray length = spray pattern width × number of spray guns = 8 in. × 4 = 32 in.

Tablet bed velocity (same as the peripheral pan velocity)

= pan circumference × pan speed = 3.14 × 67 in. × 5.0 rpm = 1,052 in./min

Tablets per minute in the spray zone:

Coating zone/min = tablet bed velocity × total spray length = 1,052 in./min × 32 in. = 33,664 in.^{2}/min

Tablets/min in the spray zone = coating zone/min × number of tablets/ in.^{2} = 33,664 in.^{2} × 22 tablets/in.^{2} = 740,608 tablets/min

Coating solution per pass through the spray zone:

$$= \frac{\text{total spray rate}}{\text{tablets/min/spray zone}} = \frac{500 \text{ g/min}}{740,608 \text{ tablets/min}}$$

= 0.675 mg/tablet (0.081 mg solids)

Passes through the spray zone:

$$\frac{\text{tablets/min/spray zone}}{\text{total number of tablets}} \times \text{spray time}$$

$$= \frac{740,608 \text{ tablets/min}}{1,700,000 \text{ total tablets}} \times 125 \text{ min} = 54.5 \text{ passes}$$

Weight gain:

weight gain/tablet = coating per pass through the spray zone × number of passes = 0.081 mg solids/pass × 54.5 passes = 4.41 mg

$$\text{total weight gain} \approx \frac{\text{weight gain/tablet}}{\text{tablet weight}}$$

$$\approx \frac{4.41 \text{ mg}}{225 \text{ mg/tablet}} \approx 2.0 \text{ percent}$$

In this analysis, each tablet would be within the spray zone approximately 55 times over the course of a 125 min coating trial. In actuality, many of these tablets would be in the spray zone either more or less frequently, depending on the uniformity of product movement. Additionally, failure to optimize the gun-to-gun delivery or spray distribution will have an adverse effect on coating uniformity. Therefore, this factor should be examined to ensure a consistently high-quality film coating.

Process Optimization

Formulation Requirements

Product Substrate (Tablet Core). To effectively develop the coating process, the product must be evaluated to ensure that it meets the necessary criteria for a substrate.

Hardness/Friability. The tablet cores must be capable of withstanding the rigors of tumbling in the coating pan. In a larger diameter coating pan with a greater load, the bed depth will be greater and, therefore, subject the tablets to greater stress. So, an acceptable tablet hardness or friability for a small coating pan may not be sufficient for a larger pan. A tablet hardness tester is used to determine the edge-to-edge (diametral) tablet hardness. Typical hardness measurement units include kp, Strong Cobb (Sc), and Newton (N). A kp is defined as the force exerted by a kilogram mass on its support in a gravitational field of g = 9.80665 m/sec^2. (1 kp = 9.807 Newton units or 1.4 Strong Cobb units.)

Tablet hardness has traditionally been the measure of a tablet's suitability for coating. However, in many cases, the tablet may be of substantial hardness but still exhibit unacceptable capping tendencies or show excessive wear on the tablet edges or logo. Therefore, a slightly better means of determining a tablet's ability to withstand tumbling is friability. This is usually determined by tumbling a certain number or weight of tablets for a set number of rotations (usually 100 revolutions) inside a cylinder. The tablets are weighed before and after tumbling and weight loss is expressed as the percent friability. A new innovation to the friability test is to line the inside of the friability cylinder with a mesh screen. This has been shown to provide a better correlation between the friability test and actual coating suitability (Chang 1994).

Weight Variation. Tablet cores are usually produced to a particular weight range specification. However, these ranges are not always narrow enough. This usually occurs when tablet granulation exhibits flow problems. A wide tablet weight variation will make it difficult or impossible to determine accurately the actual tablet weight gain due to the application of the film coating, since the weight variation in the uncoated cores can be greater than the weight of the film to be applied. Also, wide variations in tablet weight can be accompanied by variations in tablet hardness.

Stability. The tablets must be stable under the conditions required for coating. The product must be able to withstand the temperature and humidity of the process airflow. Product temperature is significantly less than that of the inlet air during the coating process due to evaporative cooling. However, during the preheat phase, the product temperature may approach that of the inlet air if the tablets are not jogged frequently enough. Tablets should be able to handle the usual product temperature of 35–50°C. As mentioned earlier, it may be necessary to use drier conditions during the initial phase of the coating process to prevent stability problems with moisture sensitive products, such as effervescent tablets.

Compatibility. The compatibility of the tablet core with the excipients in the film coating suspension must be verified. In some instances, certain actives have exhibited an interaction with the plasticizer in the coating suspension.

Shape. If possible, certain tablet shapes should be avoided for film coating. Tablets with sharp edges may exhibit a greater tendency for edge wear. Cores with a large, flat tablet face may result in poor product movement due to sliding. Also, tablets with large, flat surfaces will show a much greater tendency to exhibit twinning during coating. By adding a slight concavity (0.005 in. to 0.008 in.) to the face of the tablets, agglomeration is less likely to occur.

Logo Design. Sharp corners or small islands on the tablet logo can lead to logo attrition problems. If the logo is too fine or contains too much detail, the film coating may bridge or cover the logo. A draft angle of 35° is recommended for film-coated tablets. Tooling manufacturers are usually aware of the tool design specifications for tablets that are to be film coated.

Core Porosity. The tablet core must be formulated so that there is good adhesion between the film coating and the tablet surface. If the core porosity is low, then poor adhesion will result and picking and/or peeling of the film will occur. Core porosity or surface hydrophobicity can be a problem with wax matrix tablets due to poor adhesion between the tablet surface and film coating droplets. To remedy this, more adhesive film polymers (i.e., HPMC) may be used.

Disintegration. The disintegration of the tablet core must be rapid when tested prior to the addition of the film coating. If the core does not disintegrate quickly prior to coating, then the film coating will only delay it further.

Coating Formulations

The properties of coating formulations were discussed previously in "Guidelines/Limitations."

Process Objectives

Once all of the formulation requirements have been fulfilled, the following factors, which can be grouped into three general areas, can be examined to optimize the coating quality and coating system performance: uniformity of spray application, uniformity of product movement, and achieving an adequate evaporative rate. These factors must be examined and optimized in order to optimize the entire coating process.

Uniformity of the Spray Application. *Spray Gun Design.* There are a variety of spray gun manufacturers. Some of the more common suppliers include Binks, Freund, Schlick, and Spraying Systems (see Table 3.7). The most common type of gun used is an air atomizing (pneumatic) gun. This type of gun allows the use of variable spray rates. In the past, hydraulic gun systems were used; however, they did not allow flexibility in spray rates. With hydraulic systems, the solution nozzle orifice must be sized to match the desired spray rate. In order to change the spray rate, a change in the solution nozzle was necessary. Additionally, using a smaller nozzle orifice made these more susceptible to plugging of the gun tip.

With some types of pneumatic spray guns (e.g., those of Spraying Systems Co.), the air cap and spray nozzle configuration produce a predetermined ratio of atomization to pattern air at a given supply pressure. In other guns (e.g., those of Freund Industrial Co.), there are separate controls for adjusting the volumes of the atomization and pattern air. This allows the separate adjustment of either atomization or pattern air without having to change the air cap and/or the solution nozzle. For example, the width of the spray pattern can be changed without changing the spray droplet size. The atomization air breaks the solution stream into a fine droplet size, while the pattern air serves to flatten the spray into a fan-shaped pattern. The volume of atomization controls the mean droplet size of the spray. The pattern air volume controls the overall width of the spray.

Number of Spray Guns. The number of spray guns needs to be adequate to provide uniform coverage of the entire product bed. To maximize the uniformity and application of the coating, the spray zone should cover from the front edge to the back edge of the tablet bed. Adding more spray guns will not automatically guarantee that the overall spray rate can be increased. Using additional guns is only justified if the current number of guns is insufficient to cover the tablet bed. The objective is to produce a uniform "curtain" of spray that the tablets pass through. Spray guns are usually capable of developing pattern widths of 5 in. to 8 in., without adversely affecting the distribution of the spray droplets. At pattern widths greater than 8 in., the volume of pattern air needed to fan out the spray can lead to a distortion of the droplet size distribution due to the recombination of spray droplets. The spray guns should be set up such that adjacent spray patterns are as wide as possible without overlapping. Overlapping of spray patterns can lead to localized overwetting of the tablet bed.

Uniform Gun-to-Gun Solution Delivery. It would seem obvious that in order to achieve uniformity of suspension application, all spray guns must be set up to deliver the same quantity of coating suspension. A recent trend in coating systems is using a single pump manifolded to multiple spray guns. For these types of systems, it is mandatory that the system is calibrated on a regular basis to ensure that all spray guns are delivering the same quantity of suspension. Furthermore, calibration must be performed with the coating suspension to be used. Calibration with water will not be satisfactory, since its viscosity is much lower than the coating formulation, and it will not be as sensitive to differences in pressure drop between the spray guns. Calibration is also recommended when using peristaltic pumps, since the tubing is subject to fatigue. Calibration is usually accomplished by adjusting a knob that controls the restriction of the spray nozzle by the spray needle.

Atomization Air Volume/Droplet Size. As stated earlier, the atomization air volume can be adjusted to control the mean droplet size of the spray. Increasing the atomization air volume can reduce the mean or average droplet size. Increasing either the spray rate or solution viscosity will cause an increase in both the droplet size distribution and the mean droplet size. Therefore, the droplet size should be evaluated at the exact spray rate that will be used for the coating trial. The quality of the spray in terms of droplet size and distribution should be evaluated at several different settings.

Spray Gun Angle. Ideally, the spray gun should be directed at the middle (midway between the leading and trailing edges of the tablet bed) and at a 90° angle to the moving tablet bed. In other words, midway between the 7 and 8 o'clock position. If the spray guns are directed higher (or toward the leading edge of) of the tablet bed, it is possible that the spray could be applied to the pan or the mixing baffles as they begin to emerge from the tablet bed. Conversely, if the spray guns are directed too low on the tablet bed, spray applied to the tablets may not have sufficient time to dry. This could result in the transfer of film from the tablets to the pan surface. If the spray guns are not directed at a 90° angle to the tablet bed, the spray as it exits the solution nozzle has a tendency to build up on the wings of the air cap.

Uniformity of Product Movement. *Pan Speed.* This was discussed earlier (see under "Process Parameters"). Product movement must be uniform if a uniform application of the coating solution is to be achieved. The minimum pan speed necessary to achieve this objective is recommended. Once the pan speed has been determined, it can be scaled up by duplicating the peripheral edge speed—the ratio of the small pan diameter to the large pan diameter times the small pan speed.

One final note is that product movement should be continually evaluated and adjusted as necessary during the coating cycle. Frequently, product movement will change as a coating is applied to the tablet surface.

Tablet Size and Shape. Different tablet sizes and shapes will exhibit significantly different flow characteristics. In general, smaller tablets will flow better than larger tablets; longer, less round shapes, such as capsules and oval-shaped tablets, will tend to slide and flow more poorly than other shapes. If the tablet shape or size is changed, then the product flow properties must be reexamined. If the size is too small, a fine mesh will be required to cover the perforations in the pan to prevent the exhaust of product. Another consideration is if the product is fine (~1 mm or less). If the airflow is exhausted down through the product bed, there will be a considerable pressure drop across the bed, which will cause a reduction in the process airflow.

Baffle Type/Size/Number. The primary function of the mixing baffles is to transfer the product between the front and back of the coating pan. A variety of different baffle shapes and sizes exist. Coating pans are usually equipped with a standard baffle design that works well for most products. However, it may be necessary to use a different baffle design for unusual shapes and sizes. Coating pans can also be fitted with antislide bars, which are positioned on the flat of the coating pan and perpendicular to product flow. They are used to prevent tablets with large, flat surfaces from sliding inside the pan. With most coating pans, it is necessary to use a reduced baffle size when working with smaller batch sizes.

Standard size baffles used with a small batch will result in sluggish product movement and/or excessive variation in the spray gun–to–product distance. As the baffle passes through the product, it will temporarily carry a portion of the tablets out of the tablet bed, causing a brief increase in the gun-to-bed distance. As these tablets cascade off the baffle, the tablet bed height rises, and the gun-to-bed distance decreases. A minimum variation in the gun-to-bed distance is desired so that the spray always travels a consistent distance, and spray droplets striking the tablets have a constant moisture level. The typical variation in gun-to-bed distance is 1–2 in. A reduced or small batch baffle is generally recommended whenever the batch size is less than 75 percent of the rated pan brim volume.

Batch Size. As mentioned above, the batch size/baffle combinations are critical to obtaining acceptable product movement. An acceptable batch size range for film coating is usually from 50 to 95 percent of the rated brim volume. By using only 95 percent instead of 100 percent of the rated volume, spillage from the pan mouthring during coating can be eliminated. Occasionally, what appears to be an acceptable pan load initially may turn out to be excessive for those products that exhibit a change in product movement as a film is applied or coatings with an extremely high weight gain. A problem in working with smaller batches (~50 percent) in a perforated coating pan is that unless part of the exhaust plenum is blocked off, the process air will preferentially pass around the tablet bed due to less restriction or pressure drop. Batches of this size can be coated successfully; however, the drying efficiency is reduced.

Adequate Evaporative Rate. The theoretical evaporative rate for aqueous film coating can be determined by using the following equation:

$$\text{rate} = \frac{\text{ACFM} \times C_p \times L \times \min \times \left[(T_{\text{in}} - T_{\text{out}}) - H_L(T_{\text{in}} - T_{\text{out}})\right]}{\text{LHV}}$$

where rate is the evaporative rate in pounds of water per hour, ACFM is the actual process airflow in cubic feet per minute, C_p is the specific heat capacity of the air (Btu/lbm-°F), L is the density of the airstream (lbm of dry air/ft^3), min is the minutes per hour, T_{in} is the inlet process temperature (°F), T_{out} is the exhaust process temperature (°F), H_L is the heat loss of the coating system (percent), and LHV is the latent heat of vaporization (Btu/lbm). This equation will determine the approximate quantity of water (in pounds per hour) that can be evaporated in a coating pan system. However, it will not guarantee the quality level of the film coating. The quality level, as stated earlier, is dependent on many other factors. If each critical factor is examined and optimized, this equation can be used as a tool to determine the approximate spray rate.

For example, the following assumptions were made for a process:

- Ambient air conditions = 70°F/50 percent relative humidity.

- Air flow = 1,275 ACFM.

- Specific heat capacity of air = 0.241 (Btu/lbm-°F).

- Density of the airstream = 0.0634 (lbm of dry air/ft^3).

- Inlet air temperature = 75°C (167°F).

- Exhaust air temperature = 40°C (105°F).

- Coating solution solids = 12 percent.

- Number of spray guns = 4.

- Heat loss = 10 percent.

- Latent heat of vaporization = 1,040 (Btu/lbm).

Applying the previous equation,

$$\text{rate} = \frac{1,275 \times 0.241 \times 0.0634 \times 60 \times \left[(167° - 105°) - 0.10(167° - 105°)\right]}{1040}$$

$$= 62.71 \text{ lb/h or } 475 \text{ g/min of water evaporated}$$

At 12 percent solids, this equals 540 g/min/total or ~135 g/min/gun.

Process Air Volume. The process airstream should be adjusted to the maximum volume that yields a laminar (nonturbulent) airflow. A turbulent airflow will distort the spray patterns and lead to lower coating efficiency due to spray drying of the coating. It is important to inspect the coating pan periodically to ensure that all process air passes through the tablet bed. Any air that passes around the product will result in a lower evaporative efficiency. The inlet air volume is more important than the exhaust air volume as an indicator of the evaporative capacity, since it is this air that passes through the tablet bed and vaporizes the water from the coating and then conveys it away from the tablets. The exhaust air volume may be slightly greater than that of the inlet air due to the addition of nozzle air from the spray guns and slight leakage that may exist. If there is a large difference between the inlet and exhaust air volumes due to leakage, this can be a problem, since this air can artificially depress the exhaust temperature.

Spray Rate. The initial spray rate that is selected may be based on previous coating trials that have yielded successful results. This is an acceptable approach if the coating solution and product substrate are very similar. However, it is important to remember that changes in these factors can drastically affect the spray rate and other parameters selected. Another method for determining an initial spray rate is to evaluate the quality of the droplet size distribution. Spray rates for aqueous film coating vary from 6 to 30 g/min for a small (2.0 L) pan to 80 to 250 g/min in a large production scale pan. The key factors that limit the maximum spray rate per gun are the viscosity of the coating solution, the type of spray gun used, and the level of film quality that the customer deems acceptable.

Spray Gun-to-Tablet-Bed Distance. For small-scale coating systems, the gun-to-bed distance can be as little as 1 to 2 in. The typical gun-to-bed distance is 8 to 10 in. for a production-sized coating pan. This distance usually provides an economical trade-off between the cost of the number of guns needed to cover the spray zone adequately and the desired spray quality. If the gun-to-bed distance is less than 8 in., either the spray rate must be reduced or the inlet temperature and product temperature increased to compensate for the shortened evaporation time. If this distance is greater than 8 in., the inlet process temperature should be reduced, otherwise more spray drying will occur.

Product/Exhaust Temperature. The standard approach to the film-coating process utilizes an exhaust temperature between 38°C and 44°C. Based on the desired spray rate, an inlet temperature is determined that allows the target exhaust temperature to be maintained.

Coatings that develop greater tackiness on drying will require a higher exhaust temperature to prevent overwetting defects (i.e., the spray droplets must be slightly drier when they strike the tablet surface to prevent overwetting). Aqueous dispersions will require a slightly lower exhaust/product temperature since the polymers used in these coatings are thermoplastic or become tacky when exposed to excessive heat. A typical exhaust temperature range for these coatings is 30°C to 38°C.

The exhaust temperature is slightly lower (~1–5°C) than that of the product bed due to heat loss between the measurement points. Usually, the greater the distance between the exhaust and product temperature probes, the greater the differential. The only time when these temperatures will vary more is during preheating and cooldown. A product temperature probe will display an average of the entire tablet bed. The product temperature can be determined by using either a probe that extends into the tablet bed or an infrared temperature probe directed at the zone just above the spray zone. Tablets in the spray zone are at a slightly lower temperature due to evaporative cooling and are not representative of the average product temperature.

Dew Point Temperature. The dew point temperature is a direct measure of the moisture contained in the air. Dew point temperatures can be measured using either a capacitance sensor or a chilled, mirror-type, dew point sensor. To reproduce the drying rate from trial to trial accurately, it is recommended to maintain the dew point within a controlled range. The more critical the coating (i.e., sustained-release coatings), the tighter the range. High dew point temperatures can be controlled by using dehumidification–chilled water coils or a desiccant system. Chilled water systems are usually specified to control the dew point at 10°C to 12°C (50°F to 53°F) or an absolute moisture content of 54 to 60 grains of water per pound of air. With a desiccant dehumidification system, the dew point can routinely be controlled to as low as -6°C (21°F). Dew point temperatures can be controlled on the low end by humidifying the air via the injection of clean steam into the process air. If no attempt is made to limit the variation of the inlet dew point, then fluctuations in ambient air conditions can lead to reduced coating efficiency (spray drying), longer processing times, or film defects due to overwetting.

Film Coating Defects/Troubleshooting

Table 3.9 contains a brief definition of common tablet coating defects, along with typical causes and suggested remedies.

Table 3.9. Tablet Coating Defects and Corrective Action

OVERWETTING/PICKING: Occurs when part of the film coating is pulled off one tablet and is deposited on another. If detected early in the process, it can be corrected; if detected late in the process, the coating will probably be unacceptable.

Possible Cause	Remedies
Insufficient drying rate	Increase the inlet and exhaust temperatures.
	Increase the process air volume.
	Decrease the spray rate.
Inadequate atomization	Increase the nozzle air pressure or the atomization air volume.
Poor product movement	Increase pan speed.
	Switch to an alternative baffle design.
Poor distribution of spray	Check the uniformity of solution delivery through the spray guns.

TWINNING: A form of overwetting whereby two or more of the tablet cores are stuck together.

Possible Cause	Remedies
May be due to any of the possible causes for overwetting	
Poor tablet core design	Change tablet design to eliminate large flat surfaces.

Continued on next page.

Continued from previous page.

ORANGE PEEL: Appears as a roughened film due to spray drying. This condition relates to the level of evaporation that occurs as the spray droplets travel from the gun to the tablet bed. If there is excessive evaporation, the droplets cannot spread out and form a smooth coating. A narrow droplet size distribution is important to ensure that the majority of droplets dry at the same rate.

Possible Cause	Remedies
Excessive evaporative rate	Reduce the inlet and exhaust air temperatures.
	Reduce the gun-to-bed distance.
Excessive atomization of the spray	Reduce the nozzle air pressure or atomization air volume.
Large droplet size variation	Reduce the solution viscosity.
	Reduce the spray rate.

BRIDGING: Film coating that lifts out of the tablet logo.

Possible Cause	Remedies
Poor film adhesion	Reformulate the coating solution to improve film adhesion.
	Reformulate the core formulation to increase porosity.
Poor logo design	Redesign logo to incorporate shallower angles.

Continued on next page.

Continued from previous page.

CRACKING: In some instances, film cracking may occur due to internal stress.

Possible Cause	Remedies
Brittle film coating	Increase the addition level of the plasticizer.
	Use a different plasticizer.
	Dilute the film coating solution.
Poor film adhesion	Reduce the quantity of insoluble film coating additives.

POOR COATING UNIFORMITY: Manifested in either a visible variation in color from tablet to tablet or in the form of an unacceptable release profile for the tablets.

Possible Cause	Remedies
Insufficient coating	Apply a coating of 1.5 to 3.0 percent (for clear coating as little as 0.5 percent) weight gain to attain uniformity; requires higher levels if the color of the tablets and the film are very different.
Poor color masking	Reformulate the film coating to a darker color and/or increase the quantity of opacifier.
	Increase the pan speed.
Poor uniformity of solution application	Reduce the spray rate/increase the coating time.
	Increase the spray pattern width.
	Increase the number of spray guns.

Continued on next page.

Continued from previous page.

TABLET ATTRITION/EROSION: Exhibited when some of the product substrate exhibits a high level of friability. Signs of attrition will often be minimal or nonexistent in smaller diameter coating pans. However, when the coating process is scaled up to a production-sized pan, this problem can become more severe due to the increased batch weight and bed friction.

Possible Cause	Remedies
Insufficient tablet friability	Reformulate the tablet core to a friability of no more than 0.5 percent, with a tablet hardness of at least 6–7 kp.
Excessive pan speed	Reduce the pan speed to the minimum required to achieve a smooth and continuous bed movement.
Insufficient spray rate	Increase the spray rate to provide a protective film coating in a shorter time. This may require an adjustment in the inlet air temperature.

CORE EROSION: A type of attrition due specifically to overwetting of the tablet core. With this type of defect, excessive overwetting may cause a partial disintegration of the core surface.

Possible Cause	Remedies
Surface overwetting	Reduce the spray rate.
	Increase the inlet and exhaust air temperatures.
	Increase the spray pattern width.
	Reduce the spray droplet size/increase the atomization air volume.

Continued on next page.

Continued from previous page.

Reformulate the core with less water sensitive excipients.

Improve tablet bed movement.

Increase the bed-to-gun distance.

PEELING: Occurs when large pieces or flakes of the film coating fall off the tablet core.

Possible Cause	**Remedies**
Poor adhesion	Reduce the amount of insoluble additives in the coating solution.
	Increase the level of film former in the coating solution.
	Reformulate to incorporate a film polymer with greater adhesion.
	Reformulate the tablet core to increase porosity.
Brittle film coating	Increase and/or switch to an alternative plasticizer.

LOSS OF LOGO DEFINITION: Occurs when the tablet logo is no longer clearly legible. It may be due to one or more of the defects previously covered: core erosion, tablet attrition, or bridging. Loss of definition can also occur when the logo is filled in with spray-dried film.

Possible Cause	**Remedies**
Core erosion	Reduce the spray rate.
	Increase the inlet and exhaust air temperatures.
	Increase the spray pattern width.

Continued on next page.

Continued from previous page.

	Reduce the spray droplet size/increase the atomization air volume.
	Reformulate the core with less water sensitive excipients.
	Improve tablet bed movement.
Tablet attrition	Reformulate the tablet core to a friability of no more than 0.5 percent with a tablet hardness of at least 6–7 kp.
	Reduce the pan speed to the minimum required to achieve a smooth and continuous bed movement.
	Increase the spray rate to provide a protective film coating in a shorter time. This may require an adjustment in the inlet air temperature.
Bridging	Reformulate the coating solution to improve film adhesion.
	Reformulate the core formulation to increase porosity.
	Redesign logo to incorporate shallower angles.
Excessive drying rate	Reduce the inlet and exhaust air temperatures.
	Reduce the gun-to-bed distance.
	Reduce the nozzle air pressure or atomization air volume.

Scale-Up

Batch Size

The most accurate method of determining batch size is to load the coating pan to within 1–2 in. of the mouthring (pan opening) and then rotate the pan at the desired rpm to ensure that the pan is not overfilled. Two other methods also exist for approximating the batch size: (1) Multiply the rated brim volume by 95 percent and then multiply the resultant volume by the bulk density of the product. (2) Multiply a known ratio of batch size to pan volume for a small-scale pan by the volume of the pan being scaled to. For example,

$$\text{large pan batch size} = \frac{\text{small pan batch size}}{\text{small pan volume}} \times \text{large pan volume}$$

$$= \frac{65 \text{ kg}}{90 \text{ L}} \times 550 \text{ L} \approx 397 \text{ kg}$$

It may be necessary to make a slight adjustment in the batch size due to differences in the product movement.

Pan Speed (Angular Pan Velocity)

When scaling up a coating process, it is critical that the tablet speed through the spray zone in the larger pan is comparable to that used in the smaller pan. In other words, the pan angular velocity must be the same for both coating pans. Pan velocity can be duplicated by multiplying a ratio of the small pan diameter to large pan diameter by the pan speed used for the smaller coating pan. For example,

$$\text{pan speed for large pan} = \frac{\text{small pan diameter}}{\text{large pan diameter}} \times \text{pan speed for small pan}$$

$$= \frac{39 \text{ in. pan}}{67 \text{ in. pan}} \times 9 \text{ rpm} = 5.2 \text{ rpm}$$

This equation will yield a close estimate of the pan speed. However, subtle differences in the baffle design between coating pans may require a slight adjustment from this calculated pan speed.

Available Coating Zone

To scale up the coating process with any degree of confidence, the spray rate must be determined using the same gun-to-bed distance as in the smaller coating pan. Any change in the gun-to-bed distance will change the drying time for the spray droplets and alter the quality of the coating. The same spray rate used in the small-scale pan can also be used in the production-scale pan, assuming that the same spray gun spacing and spray pattern widths are used. Typically, using more spray guns without increasing the size of the overall spray zone will not allow an increase in either the total or per gun sprays. Increasing the total number of spray guns will only be of value if the existing spray zone is inadequately covered with fewer spray guns. If the spray pattern width used in the larger coating pan is narrower than that used in the small pan, then the spray rate per gun should be reduced in proportion to the reduction of the spray zone width. This will allow the same density of film coating to be applied per unit area of the tablet bed surface. Otherwise, the quantity of coating applied to the tablets per pass through the spray zone will increase. This will most likely change the quality of the coating and can lead to overwetting or logo bridging defects. The following is an example of scale up using a ratio of total spray zone utilized:

	Pan #1	**Pan #2**
Pan Volume	90 L	850 L
Batch size	60 kg	566 kg
Coating solution	2 kg	113 kg
Number of spray guns	2	10
Spray pattern width (per gun)	8 in.	6 in.
Spray rate (per gun)	150 g/min	to be determined
Spray rate (total)	300 g/min	to be determined
Spray time	40 min	to be determined

$$\text{spray rate for large pan} = \frac{\text{pattern width pan \#2}}{\text{pattern width pan \#1}} \times$$
$$\text{spray rate (per gun) for pan \#1}$$
$$= \frac{6 \text{ in.}}{8 \text{ in.}} \times 150 \text{ g/min/gun} = 113 \text{ g/min/gun}$$

total spray rate for pan #2 = per gun spray rate (pan #2) × number of spray guns (pan #2) =
113 g/min/gun × 10 spray guns = 1,130 g/min total

$$\text{spray time} = \frac{\text{quantity of solution to apply}}{\text{total spray rate}} = \frac{113 \text{ kg}}{1.13 \text{ kg/min}} = 100 \text{ min}$$

Spray Rate to Pan Speed Ratio

One factor that is commonly overlooked is the ratio of the spray rate to the pan speed. This ratio has serious implications for both the amount of film coating applied to the individual tablets per pass through the spray zone and the overall uniformity of the film coating itself. Therefore, with any increase in the spray rate, one should evaluate the need for increasing the pan speed. In some instances, the tablets may be able to withstand a greater application per pass through the spray zone without any adverse effects on film quality or the uniformity of the coating. Whenever both the spray rate and pan speed are increased, the tablets should be evaluated for signs of overwetting or increased tablet attrition. One problem that may result if the pan speed and spray rates are too high relative to the evaporative rate is that wet film coating from the tablets may be transferred to the pan surface.

Airflow to Spray Ratio

When scaling up the film coating process, it is recommended that the airflow used in the larger coating pan be proportional to the increase in the spray rate. If the airflow is proportionally increased as the spray rate is increased and if the spray rate per unit area of the bed surface is the same, then the same inlet and exhaust temperatures can be maintained. The inlet temperature must be increased to maintain the same evaporative rate if the airflow is not increased in the same proportion as the spray rate. If the spray rate per gun and the gun-to-bed distance are the same for both pans, the airflow can be scaled up in direct proportion to the increase in spray guns. For example,

	Pan #1	Pan #2
Spray rate (per gun)	125 g/min	125 g/min
Number of spray guns	2	4
Total spray rate	250 g/min	500 g/min
Inlet airflow	660 CFM	to be determined

$$\text{airflow for pan \#2} = \frac{\text{total spray rate pan \#2}}{\text{total spray rate pan \#1}} \times \text{airflow pan \#1}$$

$$= \frac{500 \text{ g/min}}{250 \text{ g/min}} \times 660 \text{ CFM} = 1,320 \text{ CFM}$$

AIR SUSPENSION COATING

Top Spray

The basic design of the top spray fluidized bed coater consists of three main sections (see Figure 3.22). The lowermost section, the product container, is a truncated and cone shaped. At the bottom of the product container is a fluidization screen, a finely woven mesh that retains fine powders but allows fluidization air to pass through. An air distribution plate is often used in conjunction with the fluidization screen. The air distribution plate serves to increase the pressure drop across the product bed and prevents "channeling" of the process air and aids in the consistent fluidization of the product. "Channeling" occurs when the fluidization air does not distribute evenly across the product chamber (i.e., the air takes a narrow path through the product bed), which results in areas of the product bed becoming either static or underfluidized.

Above the product container is the expansion chamber, a tall section (usually) with a larger upper diameter. The spray guns are located in this section. The function of the expansion chamber is to reduce the velocity of the process air and, thus, the fluidization of the product. Reducing the product velocity is essential to keeping material in the spray zone and out of the filters.

The third and uppermost section is the filter housing, which serves to retain the product within the processing vessel while allow the process air to be exhausted. Several different types of filter systems are available.

Process or fluidization air is generated by using an exhaust blower. The exhaust blower generates a negative static pressure within the process chamber. Air is drawn through the inlet air handling system and through the distribution plate/fluidization screen. In order to suspend or fluidize a material, the process airstream must exceed the minimum fluidization velocity for that particular product/particle size.

As the product is fluidized upward into the expansion chamber. The product passes through a spray zone where an atomized coating solution is applied. As the product rises in the expansion chamber, its velocity gradually decreases due to the increasing chamber diameter. At some point, the air velocity is no longer sufficient to fluidize the product. The product then drops down along the sidewalls of the fluid bed unit and into the bottom of the product container. This cyclical fluidization pattern is continued until the desired level of coating has been applied.

Figure 3.22. Air suspension coater. Courtesy of Vector Corporation.

Wurster Columns

The unique design of the Wurster column generates higher product velocity through the spray zone. The higher product velocity allows the use of higher spray rates. The expansion chamber and filter housing of the Wurster system are essentially the same as that of the top spray fluid bed. The expansion chamber may be slightly taller to allow the product an adequate deceleration zone. A Wurster column serves as a product container. This column contains one or more partitions (Figure 3.23). The partitions are hollow, cone-

Figure 3.23. Wurster column schematic. Courtesy of Vector Corporation.

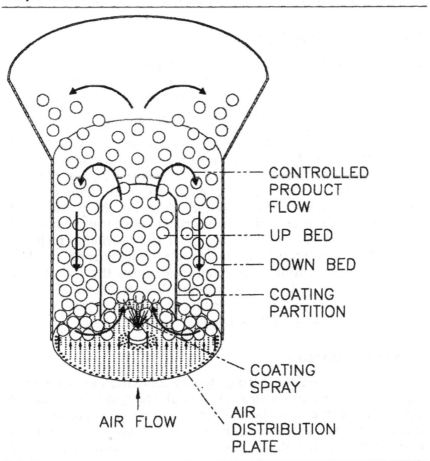

CONTROLLED PRODUCT FLOW

UP BED

DOWN BED

COATING PARTITION

COATING SPRAY

AIR DISTRIBUTION PLATE

AIR FLOW

shaped inserts mounted directly above the fluidization screen. For units with diameters up to 18 in., the partition is one-half the diameter of the column. For larger units, the partition is maintained at a 9 in. diameter. The 32 in. Wurster column has 3 partitions, and the 46 in. column has 7 partitions. Located directly below and in the center of the partition is a spray gun. The gun is directed upward into the partition. The fluidization plate is designed so that the open area around the spray gun and below the partition has a greater open area. This design causes the product to be fluidized upward through the partition at high velocity. As the product is fluidized rapidly past the spray gun, an atomized coating solution is applied. The velocity of the product lowers as the product enters the expansion chamber. The product falls back down along the outside of the partitions until it reaches the fluidization screen. The product continues this flow pattern for the duration of the coating.

Rotary Granulator/Coater

The expansion chamber and filter housing in the rotary granulator are essentially the same as those employed in the top spray or Wurster column systems. In some rotary granulator/coaters, the expansion chamber is relatively short, owing to the reduced fluidization velocities used in these systems.

Located at the bottom of the product chamber is a rotating plate or rotor (Figure 3.24). Fluidization air is supplied through a slit between the rotor and the chamber wall. In some cases, the rotor may be fitted with a screen to provide additional fluidization air. Spray guns may be mounted on either the sidewall of the product container or in the center of the chamber just above the product container. The main benefit of the rotor is that it provides a rapid and very uniform product movement through the spray zone. The high product velocity through the spray zone allows the use of spray rates comparable to those used with the Wurster column. Since the spray guns are located in the product bed, there is very little evidence of spray drying of the coatings.

System Components

A typical air suspension coating system configuration is shown in Figure 3.25. The process air goes through the system in a manner similar to the one described for coating pans. The process air is drawn through an air handler, through the air suspension unit, into a dust collector, and finally exits the exhaust blower. Figure 3.25

Figure 3.24. Rotor insert. Courtesy of Vector Corporation.

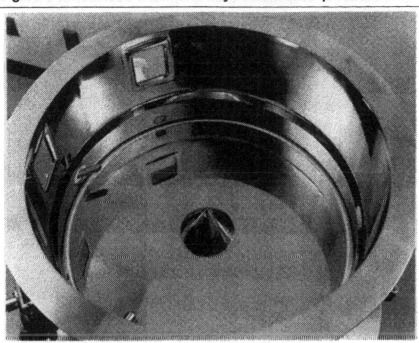

also shows the air suspension system with a bottom discharge unit into a tote. This configuration is used for large systems and the handling of toxic compounds.

Inlet Air Handling

The process air for an air suspension system is conditioned by an air handler before it arrives at the air suspension unit. The air handler is similar to the one for a coating pan except that it does not include a blower section. In rare cases, an inlet blower may be used to reduce the vacuum requirements on the exhaust blower, reduce the amount of vacuum applied to the ductwork, or fluidize a heavy product. The air should always be dehumidified when coating is done in an air suspension unit. The heater and dehumidifier are usually heat exchangers that use steam for heating and chilled water for dehumidification. Ideally, the inlet air to the air suspension unit needs to be at a constant velocity and held at a constant temperature.

Figure 3.25. Air suspension coating system. Courtesy of Vector Corporation.

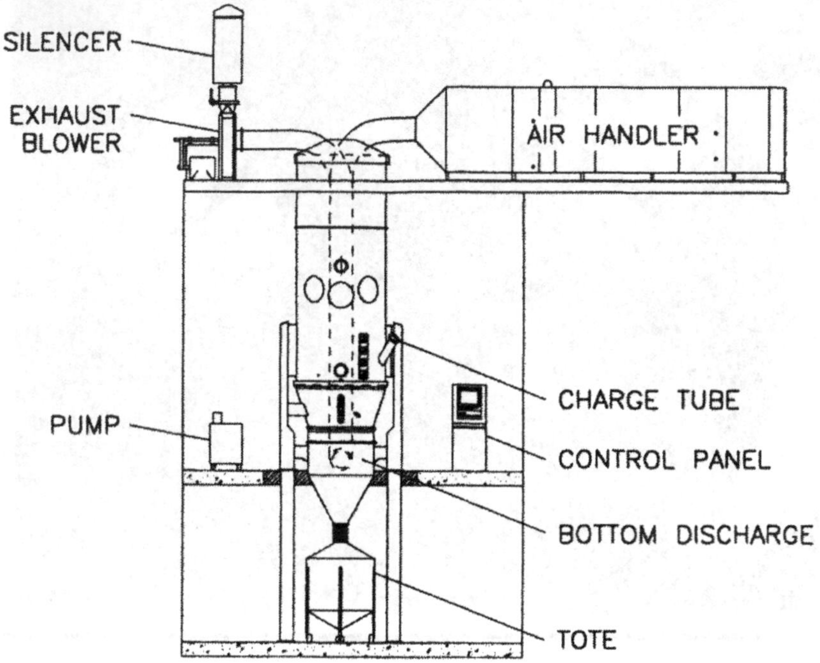

Exhaust Air System

After passing through the air suspension chamber, process air is drawn through a dust collector and into an exhaust blower. The dust collector is included as a backup for the filters in the air suspension unit and as a final filter before the process air is released to the atmosphere. If the dust collector uses cartridge filters, a bag-in–bag-out feature can be specified to prevent human contact with toxic compounds. A silencer may be placed on the exhaust blower if the blower is used in a noise-restricted area. The exhaust blower is a high-pressure blower that can develop a vacuum to overcome the pressure drop in the system and provide the required airflow to fluidize the product. This drop in pressure could be because of restrictions through the air handler, the screen slits or plates in the coating area, the product, the dust collector, the length of the ductwork, and a silencer for the blower. The blower should always be oversized by at least 50 percent, based on the calculated minimum airflow. Most manufacturers will recommend or supply the required blower for the air suspension unit sizes they provide.

Explosion Venting

Most air suspension systems are prone to create a situation where an explosion can occur. This situation is dependent on the size of the particles, the available airflow, and the potential for generating a spark. In general, the larger the particles, the less the risk of an explosion. Normally, an air suspension unit is designed with the assumption that an explosion can occur. The most common design uses a vent to release the pressure and the products of an explosion. The air suspension unit can be designed to withstand an explosion with no vents, but it must be manufactured as a 10 bar (147 psi) unit. If the unit is designed for pressures of 10 bar, the other components of the system (duct, dust collector, blowers) must be rated to withstand the same explosion.

Three components are required to create an explosion—fuel, oxygen, and an igniter. The process airflow supplies the oxygen, dust is the fuel, and static electricity is the igniter. If any of these three items is not present, an explosion cannot occur. The equipment should be grounded to a known earth ground (not a power ground) to reduce the risk of generating an ignition. All electrical equipment in the same area as the air suspension unit should be rated as explosion proof, or purged with compressed air to prevent dust from getting into the electrical components. Some manufacturers supply pneumatic controls. Any sensors or controls at the air suspension unit should be rated intrinsically safe; otherwise, pneumatic sensors and controls should be used.

All manufacturers of air suspension systems provide an explosion vent of some type to vent an explosion, if it occurs, except for units rated at 10 bar pressure. This vent may be in the side of the equipment or on the top. The side or top vent must be ducted to the atmosphere with the least amount of ductwork. In addition, the duct must be as large as the opening on the equipment and be no more than 10 ft long to prevent pressure from building up inside the equipment. The air suspension equipment should be fabricated to withstand a 2 bar (29 psi) pressure in addition to the vent. Other safeguards can be added, such as quick acting valves at the air suspension inlet and exhaust ducts to prevent the explosion from going to other equipment. Such safeguards become more important on the exhaust side because any explosion in the air suspension could trigger an explosion in the dust collector or solvent recovery unit. In addition, the dust collector and solvent recovery units should be vented.

In some cases, a suppression system can be used with the vent. Suppression units release a powder very quickly into the air suspension unit, which smothers the explosion whenever a positive pressure is detected. Ultraviolet or infrared light sensors can also be used to trigger the suppression system. To be safe, the operator should be knowledgeable about dust explosions before operating the air suspension system.

Air Suspension Filtration System

The filtration system for air suspension units is one of the key elements that allows the air suspension unit to operate. The filters in the top of the units trap particles and prevent them from being exhausted. The filters should be cleaned during the process to prevent a buildup of particles, which in turn causes a large pressure drop across the filter, thus reducing airflow. Cleaning is also performed to force smaller particles back down into the spray area. When the cleaning action occurs, the smaller particles at the top of the air suspension unit will fall back into the product circulation area and be coated. The efficiency of the air suspension coating system is increased by the filter cleaning process.

Filter Bags/Socks. The first air suspension systems were manufactured with filter bags or socks. The filter bags were generally constructed with several bags or socks to increase the filter area and were cleaned during the process by mechanically shaking the socks. Some sock filters are similar to the bags except that individual socks can be removed for replacement and cleaning. Typically, the socks use a compressed air blowback system to clean the socks during the process. The socks are mounted at the top of the air suspension units, above where the process air would fluidize the majority of the product. The exhaust duct for these units is above the filter. Smaller particles of the product will be blown into the bags or socks and retained.

When the filter material is properly sized for the product, 95–98 percent of the product can be retained in the air suspension unit. Bag filters are built in several configurations. The simplest form is a single bag with tubes hanging above the product that can be shaken by a mechanical device. The coating spray and process air must be stopped during the shaking period. The frequency of shaking is dependent on the airborne particles from the product being produced.

The bags can be constructed of several different types of material; the most common is a polyester cloth that will not pass particles of 20–50 μm or larger.

In a split bag arrangement, the bag is divided into two sections. This type of bag allows for one half of the filter to be shaken while the other half passes the process air. A more common type of split bag arrangement uses a compressed air blowback nozzle in each half of the bag. Compressed air blows in a direction opposite to that of the process air into one side of the bag to clean the tubes or socks, while the other half of the bag allows the process air to flow through. In this system, the process does not have to stop to clean the filters.

A system where the spray and airflow do not have to be stopped decreases the process time by 25–30 percent. Most air suspension manufacturers offer both types of units. The mechanical bag filter should be used if the product is going to be wet and can cause caking on the filter; for most coating processes, however, this will not be the case.

Cartridge Filters. Air suspension manufacturers have recently begun to offer units with cartridge filters in place of bag filters. Figure 3.26 shows the design of a cartridge filter system. Such a system also uses a compressed air blowback system. Compressed air is injected through some of the cartridge filters in the opposite direction of the process air. The remaining cartridge filters allow the process air to flow through the product. The cartridge filters are manufactured in such a way that a maximum area can be exposed to the product. Figure 3.27 shows a cartridge filter unit; note the pleating of the material to increase the filter area. An advantage that the cartridge filter offers is that the explosion vent can now be positioned on the top of the unit—the explosion (if it occurs) does not have to go through the filter. Using top-vented units is desirable because side-vented units are prone to leaking. Changing of the filter can be accomplished easily, and a bag-in–bag-out system can be installed. These filters can be sized to stop 2 to 100 μm particles or larger at a 99 percent efficiency (99 percent of the particles larger than the specified particle size will not pass through the filter). Cartridge filters can be installed so that the filters can be removed without human contact when toxic compounds are processed.

Filter Screens. A filter screen can be installed in place of filters when the product is large, such as tablets. Such a method may also be

Figure 3.26. Cartridge filter operation in an air suspension unit. Courtesy of Vector Corporation.

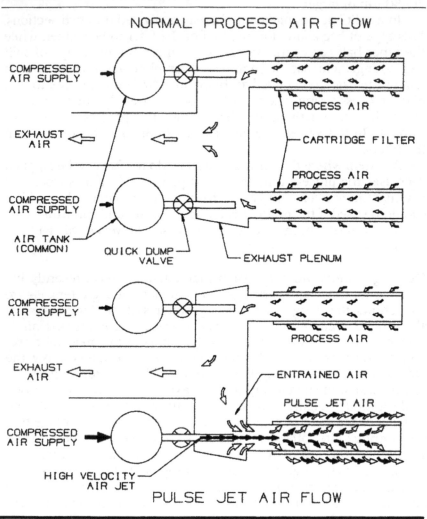

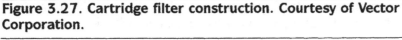

Figure 3.27. Cartridge filter construction. Courtesy of Vector Corporation.

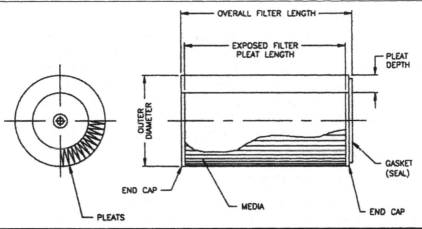

used when a product has a very narrow product size distribution; a screen can be installed that has smaller openings than the product size. This approach reduces the pressure drop across the filter area, allowing higher airflows and better fluidization for the larger product size.

Spray Systems

Spray systems for the air suspension units are very similar to the spray systems described for the coating pans; the same types of spray guns and pumps can be used. However, the pumps for the air suspension systems are generally rated for lower flow rates than those for coatings pans; thus, peristaltic pumps can be used. In larger air suspension units, a single positive displacement pump may be used to feed all of the spray heads. The spray system of choice is pneumatic guns in a closed recirculation loop configuration, where a separate pump feeds each spray gun, and the solution is recirculated when there is no spray.

Spray Guns. Air suspension spray guns are mounted in one of three configurations. They can be mounted on the periphery of the air suspension unit, inside the unit above the product, or as an up-spray unit. When the Wurster column is used, the spray gun is mounted at the bottom of the Wurster insert as an up-spray unit.

Three manufacturers of spray guns used in air suspension units are Vector Corp. (Freund), Spraying Systems, and Schlick.

Instrumentation Control

The system controls for air suspension coating can be similar to the controls described for coating pans. However, RPM is not required because there is no rotating drum. The static pressure in the lower chamber, expansion chamber, and after the top unit filters should be held constant. This is accomplished by measuring and controlling the exhaust blower similar to that used for the coating pans. Other parameters described for the coating pans will provide similar analog and digital signals.

Manufacturer Comparisons

Four major manufacturers in the United States provide most of the air suspension units that are used for coating. Most of these units can serve as dryers or granulators; with some options, they can be used for other processes. The four manufacturers, listed in Table 3.10, build units that range from lab size models to very large processing units. The lab size model is used to test a process and is then scaled up to a production size unit. All of the units can coat products as long as the process is properly developed. When one considers buying an air suspension unit, the complete system must be considered. The manufacturers listed in Table 3.10 will supply a complete system, and most will offer a turnkey option for complete installation, including services for performing the installation qualification (IQ) and operational qualifications (OQ) when validation is required. The following basic items should be considered before choosing a system:

- Dedicated unit or a multipurpose unit (Wurster insert, rotor insert).

- Capacity of the air suspension unit (this determines the size of the blower, air handler, and dust collector).

- Cost of the system.

- Options available (type of filter, product cart or bottom discharge, type of pump, dehumidification, humidification, face and bypass air handler, etc.).

- Type of control system (manual, semiautomatic, automatic).

- Location of explosion venting (side or top).

Table 3.10. Air Suspension Manufacturers

Manufacturer	Address	Manufacturing Location
Aeromatic, Inc. (Niro Inc.)	Columbia, Maryland	United States and Denmark
Fluid Air Inc.	Aurora, Illinois	United States
Glatt Air Techniques Inc.	Ramsey, New Jersey	Germany
Vector Corporation (Freund)	Marion, Iowa	United States and Japan

- Location to be installed (this will determine the length of ductwork, size of the blower, and the type of explosion venting).

- Type of spray system (down, up, tangential, or peripheral).

- Level of explosion requirements (venting only, barrier valves, suppression).

- Type of pump desired (peristaltic, gear, lobe).

- Type of filters (bag with mechanical, bag with blowback, cartridge).

In general, the manufacturers given in Table 3.10 will provide special options that may be required for the user's application. Table 3.11 lists the specifications for general, production-sized units. Other options, such as bottom discharge units, automatic loading ports, clean-in-place units, and jacketed solution tanks, can be provided.

Individual air suspension systems can be designed, but it is highly recommended that the manufacturer provide the complete system because they know the limitations of their equipment, and, in many cases, the problems that can be encountered with different products having different characteristics.

Table 3.11. Production Air Suspension Models and Specifications

Manufacturer	Model	Width (inches)	Height (inches)	Capacity (Liters)
Fluid Air	0500	86	75	610
Fluid Air	0800	96	75	950
Fluid Air	1000	105	75	1,180
Fluid Air	1200	114	75	1,450
Glatt Air Techniques Inc.	120	59	191	420
Glatt Air Techniques Inc.	200	71	199	670
Glatt Air Techniques Inc.	300	71	206	1,100
Glatt Air Techniques Inc.	500	106	269	1,560
Niro Inc. (Aeromatic, Inc.)	MP-5	43	157	480
Niro Inc. (Aeromatic, Inc.)	MP-6	51	171	780
Niro Inc. (Aeromatic, Inc.)	MP-7	59	188	1,085
Niro Inc. (Aeromatic, Inc.)	MP-8	67	203	1,525
Vector Corporation	FL-N 60	55	120	220
Vector Corporation	FL-N 120	69	126	420
Vector Corporation	FL-N 200	76	151	660
Vector Corporation	FL-N 300	79	170	1,100

Process Parameters/Coating Dynamics

Many of the same factors that are critical to film coating in a coating pan are also relevant in a fluid bed system. These include the quality of the product movement, proper evaporative rate, and a uniform distribution of the spray onto the product.

Airflow

Air Velocity/Volume. Process air in the fluid bed not only generates the evaporative capacity for the unit but also provides the necessary product movement. Product movement is the determinant for optimizing the fluidization air volume or velocity. The fluidization air volume must be selected such that the product exhibits an adequate velocity through the spray zone. Yet the velocity must be low enough to prevent the product from being overfluidized or trapped in the filtration system. In other words, the product must exceed minimum fluidization velocity in the product container but be lower than this velocity at the filter housing. Additionally, the entire product bed should be in constant motion. It is imperative that product movement be monitored on a regular basis since any change in the product's size or density will change the optimal fluidization volume.

Minimum Fluidization Gas Velocity. The minimum fluidization gas velocity is dependent on the material's size and density. The required minimum fluidization gas velocity can be calculated as follows (Boucher 1973):

$$G_{mf} = 688 \frac{\left[p_f(p_s - p_f) \right]^{0.94}}{\mu^{0.88}} d_p^{1.82}$$

where G_{mf} is the minimum fluidization gas velocity [lb/h(ft^2)], μ is the fluid viscosity (0.018 cps for air), p_f is the fluid density (0.075 lb/ft^2), p_s is the true density of the fluidized particle (in lb/ft^3), and d_p is the particle diameter (in inches).

If the material to be processed is a mixture of materials with vastly different particle sizes or densities, the fluidization velocity must be computed for each separate component. If one material requires a minimum fluidization velocity that exceeds the terminal velocity for another component of the blend, then it is unlikely that uniform product fluidization can be achieved. Therefore, when processing a blend of different materials, it is important, whenever possible, to select materials with a similar particle size distribution and density. As a starting point, an air volume that yields 5 to 6 times the computed minimum fluidization velocity is used.

For a given product, the optimum fluidization air velocity is constant, regardless of the batch size. However, the air volume must change in proportion to the increase in bed weight. The pressure drop across the product bed can be determined as follows:

$$\Delta P = \frac{W}{S}$$

here ΔP is the pressure drop (in kg/m^2), W is the batch weight (in kg), and S is the cross-sectional area of the screened inlet opening (in m^2).

To fluidize a 50 percent batch of particles, the pressure drop across the bed will be 50 percent of the pressure drop across a full batch. Additionally, the initial or resting bed height of the half batch is approximately 50 percent of that for a full batch. This can create difficulties in that the smaller batch must be fluidized at a higher rate to achieve optimal speed through the spray zone. However, increasing the fluidization height requires increasing the fluidization air velocity beyond the optimal level, which can lead to slugging and/or excessive loading of the exhaust filters.

Rotor Gap (Rotary Fluid Bed Only). With some rotary fluid bed units, the gap between the rotor and the sidewall (see Figure 3.24) can be increased. However, increasing the gap width results in a reduction in the air velocity and a pressure drop at the rotor gap. Moreover, it allows the fluidization air volume to be increased. Typical gap distances vary from 1 to 5 mm. It is important to maintain the product above the minimum fluidization air velocity in a rotary fluid bed to prevent material from dropping into the rotor gap.

Rotor Speed/Velocity (Rotary Fluid Bed Only). The rotor in a rotary fluid bed has two important functions: (1) improve the uniformity of the product flow and (2) increase the mechanical shear of the product bed. The second function is of primary benefit when producing granulations, while the first function is important for both granulations and coatings.

When using the rotary fluid bed for coating, an increase in the rotor speed will increase the product speed (up to a certain point). At extremely high speeds, an increase in the rotor speed may result in little or no further increase in product velocity due to slippage between the product and rotor surface. However, it is not usually recommended to use these velocities, since it is probable that some product attrition will result.

The rotor speed should be selected such that the product velocity through the spray zone is rapid, without leading to product attrition. The optimal product velocity will be dependent on the type and size of material coated as well as the type of rotor and the fluidization air volume. If signs of product slippage are present, it is usually a sign that the rotor velocity is too high.

Humidity

As with a coating pan, the humidity of the fluidization airstream will directly affect the evaporative capacity and drying rate. The lower the humidity (at a given temperature) or dew point, the greater the moisture-carrying capacity of the air. If the humidity of the inlet airstream is too great, the spray rate that can be achieved without overwetting the product is limited. For critical coatings (i.e., sustained released coatings), it is important to maintain a consistent film quality, which can only be achieved with a consistent evaporative rate. With these types of coatings, it is best to control both the high and low limits of the humidity; therefore, using both dehumidification and humidification are recommended. If the inlet humidity is too low, problems may result due to static buildup. In this case, it may be necessary to humidify the inlet air to aid in the dissipation of the static charge.

Temperatures

Control by Inlet, Exhaust, or Product Temperatures. The standard approach is temperature control based on the inlet temperature. Quite often, the inlet temperature is dictated by the type of coating solution used. If the coating is an aqueous-based solution, the inlet temperature will usually be between 60° and 100°C. Coating solutions that use organic solvents vaporize much more rapidly and will use a lower inlet temperature, usually from 30° to 50°C. The inlet temperature is indicative of the evaporative capacity of the process and, therefore, is of primary importance. For a given spray, there will be a direct correlation between a given inlet temperature and exhaust/product temperature. The exhaust and product temperatures will tend to follow each other closely, with the exhaust being slightly lower due to heat loss between the measurement points. However, some claim that it is more difficult to tune the exhaust or product temperature loops due to the lag time in the stabilization of these temperatures (Campbell 1997). For control based on the inlet temperature, the product or exhaust temperature is usually increased to a point slightly higher than that desired for the coating trial. This is designed to compensate for the slight drop in the exhaust or product temperature that will occur due to evaporative cooling of the product. With control based on either the exhaust or product temperature, coating can begin when the desired temperature is reached, since the inlet temperature will be adjusted to maintain these temperatures. In some cases, the control is based on a combination of the inlet and product temperatures. In some systems, the inlet temperature is controlled to preheat the product to a desired transition temperature, then the system switches to product temperature control. This type of control is useful for coating materials that have a narrow acceptable range for product temperature. For example, hot melt coating will exhibit premature congealing of the wax and produce a rough coating if the product temperature is too low. On the other hand, if the temperature is too high, the coating will become too soft, and agglomeration of the product will result.

Spray Application

Spray Gun Position. In rotary and top spray fluid bed units, there are usually one or more ports for adjusting the spray gun height. The spray gun position is usually determined on the basis of the batch size. The larger the batch size, the higher the spray gun position. The goal is to position the gun so that product is rapidly fluidized beyond the spray gun, while decelerating in the expansion chamber before reaching the exhaust filters. If the product is allowed to decelerate in the spray zone, then it is possible that the product will pick up excessive spray per pass through the spray zone.

Spray Rate. The most common method of determining the proper spray rate is trial and error. The initial spray rate is usually a spray rate that has been used successfully with a similar product. However, since there are subtle differences between the coating solutions and product substrates, it is necessary to conduct additional trials to fine-tune the coating process. Some general rules used to determine the optimal spray rate are as follows:

- The tackier the coating on drying, the slower the spray rate to prevent agglomeration of the product.

- The more viscous the coating, the slower the spray rate.

- The larger the spray zone, the greater the spray rate.

Atomization Volume/Pressure. The atomization air volume must be adequate to produce a uniformly fine droplet size distribution. The greater the atomization air volume or pressure (all other things being equal), the finer the mean droplet size of the spray. The smaller the particle to be coated, the finer the droplet must be to prevent agglomeration. For extremely fine particles, the droplet size must not be greater than one-tenth the size of the substrate particle. As the viscosity of the coating solution is increased, the atomization air volume must also be increased to prevent larger droplet sizes. An increase in the spray rate will also require an increase in the atomization air volume to maintain the same mean droplet size.

Atomization Air Temperature. For certain coating materials, most notably hot melt coatings, it is necessary to heat the atomization air to prevent premature congealing of the spray. The temperature of the air should be at least as great as the melting point of the hot melt material. This heated atomization air also serves to heat the spray gun, thus preventing solidification of the hot melt inside the gun itself.

Droplet Size Distribution. The quality of the spray (i.e., the droplet size distribution) can be a major factor in determining the maximum acceptable spray rate per spray gun. A narrow droplet size distribution results in less variation in the residual moisture content within the spray droplet as it strikes the product surface. A more consistent spray droplet moisture content allows the droplets to dry at a more uniform rate. A more uniform drying rate will translate into a smoother film surface and a higher coating efficiency (less spray drying). Increasing the spray rate or solution viscosity not only increases the droplet size but also increases the droplet size distribution. Increasing the atomization air volume will compensate for minor changes in spray rate or solution viscosity; however, major changes will ultimately have some effect on quality or coating efficiency.

Coating Zone/Pattern

Number and Type of Spray Guns. The overall size of the coating zone is ultimately the limiting factor for the overall spray rate. If the fluid bed unit was infinitely large and could accommodate an infinite number of spray guns, the total spray rate would be unlimited. For film coating in a top spray or rotary fluid bed, the more guns that can be used without overlapping adjacent spray patterns or spraying onto the sidewall of the spray chamber, the greater the overall spray rate. In the Wurster column, the more partitions/spray guns that can be used, the greater the overall spray rate. Additionally, the size of the spray zone per spray gun is a major factor in determining the maximum spray rate per spray gun. Even if the fluid bed system has the additional evaporative capacity, there is a limit to how much coating can be applied to the product surface per pass through the spray zone before uniformity or quality begins to deteriorate. If the spray gun produces a larger zone over which to distribute the spray, then the spray rate for that spray gun can be greater. This assumes that the quality of the spray (i.e., the droplet size distribution) is maintained.

Comparison of Guns from Different Manufacturers. The most common brand of spray gun currently used in fluid bed systems is Schlick®. Schlick® guns are used for both top spray and Wurster column applications. These guns utilize a fixed, cone-shaped pattern. The droplet size is controlled by adjusting the spray air pressure. The spray gun for the Wurster column usually has a narrower and longer gun body, which enables this gun to extend into the partition without obstructing the flow of product. The top spray Schlick® spray guns have multiple nozzles. The most common being those containing three or six nozzles. Some of the earlier versions of the top spray fluid beds utilized peripherally mounted guns located in the expansion chamber wall. These Freund type guns differ not only in their location but also in that they employ a fan-shaped pattern. These guns have separate controls for adjusting the droplet size and the width of the spray pattern. The Freund spray gun provides excellent coverage of the product bed; however, they are slightly more difficult to calibrate due to the increased number of spray guns used.

Spray guns used for rotary fluid bed systems are either peripherally or tangentially mounted. These spray guns are located in the sidewall of the product container or submerged in the product during coating. The spray pattern is cone shaped. The spray gun is angled such that solution application is concurrent with product flow.

The spray gun system may be either recirculating or nonrecirculating. In the recirculation system, the solution flows through the spray gun (or through some part of the inlet solution line) and then back to the solution tank (refer to Figures 3.18a and 3.18b). The initiation of the spray activates the spray needle, which is retracted from the solution nozzle. At the same time, a stop valve (either inside the gun body or in the solution return line) is activated, which prevents flow back to the solution tank. At this point, the solution must exit through the solution nozzle.

The spray gun is usually either a coaxial or triaxial design. With coaxial design, the spray wand or bar contains two separate tubes for delivering the solution and the nozzle air volumes. The triaxial design contains a third tube for the spray air, which activates a spray needle to start or stop the flow of solution through the solution nozzle.

Coating Analysis

One of the most important rules to remember when developing a coating process in any piece of equipment is that virtually all parameters are interdependent. A change in one parameter will have an effect on other operating parameters. For example, a change in the spray rate may require an adjustment in the atomization pressure or inlet temperature.

The most important factors or parameters to examine when optimizing the coating process are as follows:

- Product movement.

- Spray rate.

- Inlet temperature.

- Atomization air pressure.

For any fluid bed application, product movement through the spray zone must be vigorous. Sluggish product movement through the spray zone will lead to either poor coating uniformity or a drastic reduction in the achievable spray rate.

The spray rate per spray gun is limited by the size of the spray zone and the product velocity through that zone. In other words, only so much coating can be applied to the product substrate per pass through the spray zone. An increase in airflow or inlet temperature will increase the evaporative capacity of the unit, but it will not allow a higher spray rate without an adverse effect on the film coating. It is essential that the proper evaporative rate is used to support the desired spray rate. The airflow is limited to that needed to fluidize the product adequately. Therefore, the inlet temperature is used to compensate for increases or decreases in the spray rate.

Process Optimization

Core Specifications

Essentially, the same requirements mentioned earlier for film coating in a pan hold true for coating in a fluid bed coater. However, a product that might be suitable for coating in a pan may not be suitable for a fluid bed. For example, the vigorous fluidization that occurs in a fluid bed may require a core substrate of greater hardness than that in a coating pan (assuming the same size substrate). Also, a product shape that flows well in a coating pan may not exhibit acceptable flow in a fluid bed coater. The fluid bed coater is capable of coating much smaller particles than the coating pan. Product passes through the spray zone at a higher velocity than in a coating pan. Moreover, in the fluid bed, the product is not in intimate contact in the spray zone like it is in the coating pan, which means it is less susceptible to agglomeration. There is a limit to how fine a product can be coated in the fluid bed without producing agglomeration. This limit is usually somewhere between 100 and 150 μm; however, it is highly dependent on the coating formulation used (Sackett 1997).

Coating Solution

The requirements for a coating solution in a coating pan are comparable to those for fluid bed coating. Since the fluid bed is often used to coat smaller particles, the limitation on the solution viscosity is more restrictive. A solution that may work well in a coating pan may be difficult or impossible to use for fine particle coating in the fluid bed without producing product agglomeration. For fine particles, a coating solution that is of low viscosity and tackiness is preferred.

Process Objectives

Uniformity of Spray Application

To achieve a high quality coating, it is essential to apply the film coating uniformly. All of the spray guns must deliver coating solution at the same rate, and product movement through the spray zones must be of a sufficient velocity. Routine calibration of the spray guns must be conducted with the actual coating solution. Additionally, the fluidization air volume must be adjusted such that the product movement is consistent through the different spray zones. Typically, it is necessary to add a distribution plate beneath the fluidization screen to prevent "channeling" or nonuniform product movement. The distribution plate eliminates or minimizes inconsistent product movement by increasing the pressure drop across the product bed and increasing the face velocity at the bottom of the product container.

Uniform Product Movement

For rotary fluid bed units, the product velocity can be optimized by adjusting the rotor speed in conjunction with the air volume. In the Wurster column, product movement is optimized by selecting the proper air distribution plate. The larger the product size, the greater the open area in the down-bed zone (that portion of the distribution plate not beneath the partition); see Figure 3.23. Hence, selecting the appropriate air distribution plate allows larger product sizes to maintain proper fluidization in the down-bed zone. Additionally, the height of the partition must be adjusted for the proper combination of product and airflow through the partition. If the partition is set too low, then the product density through the partition will be too low. If the partition is too high, then air from the up-bed will be diverted to outside the partition, which leads to erratic product movement at the base of the partition. Usually, the larger the product size, the higher the partition to provide adequate transition of product from the down-bed to the up-bed zones (the area beneath the partitions). The air volume in a Wurster column should be adjusted so that it exits the partition at a high velocity, while decelerating before reaching the filter housing.

Adequate Evaporative Rate

Product/Exhaust Temperatures. Usually, there is a close correlation between an adequate drying rate and the exhaust or product temperatures. The product provides a more rapid indication of the evaporative rate, since the exhaust temperature lags behind that of the product. The more consistent the incoming air in terms of absolute moisture content or dew point, the better the correlation between product temperature and evaporative rate. If the product is maintained at the same temperature, but the moisture in the inlet airstream varies, then the evaporative rate of the coating process will also vary. The proper evaporative rate is determined after the uniformity of the spray application and product movement has been verified. This means that the proper inlet air temperature is selected based on the desired spray rate.

Dew Point Control. The dew point temperature is a direct indication of the absolute quantity of moisture in the air. Therefore, the best method of ensuring that the process is consistent from one coating to the next is to condition the inlet air to the same dew point. An inlet air handler that has the capability to both humidify and dehumidify the air can be controlled to maintain a single dew point. This type of system is recommended for highly critical coatings. A lesser level of control only utilizes dehumidification. This will ensure that on humid days the incoming process air will not exceed a specific dew point (typically 45° to 50°C). However, this will not prevent the coating from drying too rapidly on those days when the ambient dew point is extremely low.

Troubleshooting: Problems (Equipment and Operational Sources) and Corrective Action

Table 3.12 describes some common fluid bed coating defects along with typical causes and suggested remedies.

Table 3.12. Film Coating Defects/Corrective Action

OVERWETTING/AGGLOMERATION: Occurs when insufficient drying rate conditions exist. For large product substrates, such as tablets, overwetting may be manifested as picking of the film. For smaller particles, overwetting usually results in irreversible agglomeration.

Possible Cause	Remedies
Insufficient drying rate	Increase the inlet temperature.
	Decrease the spray rate.
Inadequate atomization	Increase the nozzle air pressure or the atomization air volume.
	Check for obstructions or excessive pressure drops in the nozzle air lines.
Poor product movement	Increase the process air volume.
	Install a distribution plate (or use a plate with a different percentage of open area).
	Adjust the height of partitions (for Wurster columns).
	Adjust the rotor speed (for rotary fluid bed coaters).
Poor spray distribution	Check the uniformity of solution delivery through the spray guns.
Poor solution formulation	Dilute the coating solution to a lower viscosity.
	Add an antitack agent (i.e., talc or titanium dioxide) to the coating.
	Use an alternative film polymer that has a reduced tackiness.
Improper core size/shape	Increase the size of the core substrate.
	Use a product of an alternative shape.

Continued on next page.

Continued from previous page.

SPRAY DRYING: Relates to the level of evaporation that occurs as the spray droplets travel from the spray gun to the product surface. If there is excessive evaporation, the droplets do not have the ability to spread and form a smooth coating. This condition is more pronounced in top spray fluid beds, where the spray direction is opposite to product flow.

Possible Cause	Remedies
Excessive evaporative rate	Reduce the inlet, product, and exhaust air temperatures.
	Reduce the height of the spray gun (top spray systems).
Excessive atomization of the spray	Reduce the nozzle air pressure or atomization air volume.
Large droplet size variation	Reduce solution viscosity.

CRACKING: Film cracking may occur due to internal stress.

Possible Cause	Remedies
Brittle film coating	Increase the level of the plasticizer.
	Use a different plasticizer.
Poor film adhesion	Dilute the film coating solution.
	Reduce the quantity of insoluble film coating additives.

Continued on next page.

Continued from previous page.

POOR COATING UNIFORMITY: Manifested in either a visible variation in the coating or in an unacceptable release profile for the product.

Possible Cause	**Remedies**
Insufficient coating	Increase the quantity of coating required to achieve a functional coating increases due to an increase in the total surface area (compared to larger particles). Typically, the level of coating required to achieve either an enteric or sustained release is equal to or greater than a 10 percent weight gain.
Poor uniformity of solution application	Increase the number of spray guns (if they can be added without overlapping the spray zones).
	Improve the quality of the product movement.
	Reduce the spray rate/increase the coating time.

CORE EROSION: Exhibited when some portion of the product substrate exhibits a high level of friability. As in coating pans, core erosion may not be exhibited until scaled to production-sized equipment. Erosion may also occur due to overwetting, which causes a partial disintegration of the core surface.

Possible Cause	**Remedies**
Insufficient friability	Select a new substrate or modify the existing product.
Excessive fluidization	Reduce fluidization to the minimum required to achieve a vigorous bed movement.

Continued on next page.

Continued from previous page.

Possible Cause	Remedies
Insufficient spray rate	Increase the spray rate to provide a protective film coating in a shorter time, which may require an adjustment in the inlet air temperature.
Surface overwetting	Reduce the spray rate.
	Increase the inlet, product, and exhaust air temperatures.
	Reduce the spray droplet size/increase the atomization air volume.
	Reformulate the core with less water sensitive excipients.
	Improve product movement.

BLOCKING: The agglomeration or seizing of the product after it has been coated and discharged from the fluid bed.

Possible Cause	Remedies
Inadequate drying or excess residual product moisture	Reduce the spray rate or increase the process temperatures.
	Add a dry cycle at the end of coating.
Temperature sensitive coatings (thermoplastic film polymers)	Add a cool down cycle after the coating cycle.
	Add an antitack agent prior to discharge of the product.
Excessive plasticization of the film polymer	Reduce the level of the plasticizer in the coating solution.

Continued on next page.

Continued from previous page.

POOR FLUIDIZATION/PRODUCT MOVEMENT: To optimize coating uniformity, the product must move rapidly and consistently through the spray zone, otherwise variations in coating thickness will result.

Possible Cause	Remedies
Improper distribution plate	Use a distribution plate with less open area to reduce channeling of airflow.
Insufficient airflow	Increase the airflow by opening the exhaust damper.
	Increase the slit opening (rotor fluid bed).
	Verify that the exhaust filters are not plugged.
	Verify that the fluidization screen is clean.
	Check the pressure drop across the dust collector.
	Check for obstructions in the inlet or exhaust ducts.
Poor transfer of product through the spray zone	Adjust the partition height (Wurster column).
	Increase the rotor speed (rotor fluid bed).
Static electricity	Humidify the inlet airstream.
	Begin spraying as soon as possible.
Batch size too large	Reduce the batch size.

Scale-Up

Batch Size

Fluid bed coaters have a specification for container volume. For top spray applications, the preferred batch size is usually 60 to 90 percent of the specified container capacity. In some cases, this range is expanded from as little as 35 percent to as much as 100 percent. For the rotary fluid bed coater, this range can be 20 to 100 percent of the specified unit volume. When using an extremely small or large batch size, it is essential that adequate product movement be achieved. If a small batch is used and erratic bed movement cannot be overcome, the batch size must be increased. For large batches, uneven or excessive fill at the end of coating may lead to product loss due to spillage. For top spray or rotary fluid bed coaters, the batch size can be scaled to a larger unit by using the ratio of the unit volumes multiplied by the batch size in the smaller unit:

$$B_L = \frac{V_L}{V_S} \times B_S$$

where V_L is the volume of the larger fluid bed coater (assume 1,100 L), V_S is the volume of the smaller fluid bed coater (assume 220 L), B_S is the batch size used in the smaller fluid bed coater (assume 100 kg), and B_L is the batch size to be used in the larger fluid bed coater. For example,

$$500 \text{ kg} = \frac{1,100 \text{ L}}{220 \text{ L}} \times 100 \text{ kg}$$

The maximum batch size in a Wurster column is the container volume minus the volume of the partitions. The minimum batch size is one-half of the maximum batch size. For coating in the Wurster column, the following formulas can be used to determine the minimum and maximum recommended batch sizes:

$$\text{maximum batch size} = D_B \times \left[V_C - (N_P \times V_P) \right]$$
$$\text{minimum batch size} = 0.5 \left\{ D_B \times \left[V_C - (N_P \times V_P) \right] \right\}$$

where D_B is bulk density of the product (in g/cm^3), V_C is the rated volume of the column (in L), N_P is the number of partitions, and V_P is the volume of the partitions (in L).

Fluidization Air Velocity

The product movement in a fluid bed can usually be duplicated in a larger unit by using the same fluidization air velocity. One approach is to use the cross-sectional area at the fluidization screen to determine the velocities for scale-up to a larger fluid bed. However, it may be more appropriate to use the cross-sectional area at the height of the spray guns, since it is the product velocity at this point that is most important. The air velocity is usually calculated based on the air volume and the cross-sectional area, since most fluid bed systems monitor air volume, not velocity. The air velocity in the smaller unit is then used to calculate the air volume needed in the larger unit. Regardless of which areas are used, this approach yields an initial starting point for determining the fluidization air velocity and volume. The actual product movement should always be examined and adjusted as necessary. The following equations can be used to scale-up the fluidization air velocity:

Scale-up from a 20 L to a 1,100 L fluid bed

	20 L unit	1,100 L unit
Cross-sectional area	1.35 ft²	21.5 ft²
Air volume	90 CFM	to be determined

$$\text{velocity (small fluid bed)} = \frac{\text{air volume}}{\text{cross - sectional area} \times 60 \text{ sec}}$$

$$= \frac{90 \text{ CFM}}{1.35 \text{ ft}^2 \times 60 \text{ sec}} = 1.11 \text{ ft/sec}$$

Air volume (large bed) = velocity (small bed) × cross-sectional area (large bed) × 60 sec = 1.11 ft/sec × 21.5 ft² × 60 sec = 1433 CFM

Fluidization Plate (Wurster Column). The percent open areas in the down-bed section of the air distribution or fluidization plates varies depending on the product being coated. Small particles usually require less air volume in the down-bed zone and, therefore, use a smaller percent open area. Additionally, for scaling to the 18 in. and larger Wurster columns, there is also an increase in the percent open area in the down-bed zone to compensate for the greater bed depth. The open area under the up-bed zone is usually adjustable so that the proper balance between these two zones can be achieved.

Rotor Speed (Rotary Fluid Bed)

Scale-up of the rotor speed between two rotary fluid bed units is accomplished by using the ratios of the rotor diameters. This will yield a similar peripheral edge velocity for both units.

$$R_L = R_S \times \frac{D_S}{D_L}$$

where R_L is the rotor speed for the large bed (in rpm), R_S is the rotor speed for the small bed (in rpm), D_S is the rotor diameter for the small bed (in inches), and D_L is the rotor diameter for the large bed (in inches).

Spray Rate

For fluid bed granulation, spray rates are scaled up in direct proportion to the increase in the fluidization air volume. This makes sense because the goal for fluid bed granulation is to achieve a critical bed moisture content. The goal for fluid bed coating is the same as film coating in the coating pan–to achieve a uniform application of coating to the product surface. Consequently, the most appropriate method for scaling up the spray rate is the available spray zone. If the spray guns used in the small- and large-scale fluid bed units are the same type, with the same number of nozzles and the same spacing, the ratio of the number of spray guns can be used to determine the usable spray rate. This applies to top spray, Wurster column, and rotary fluid bed coating systems. If the guns in the two units produce spray patterns of different diameter, the spray rate per gun should be scaled in relation to the area of the respective spray patterns. However, if the spray rate per gun is calculated to be substantially greater (~25 percent or more), it is advisable to verify that the droplet size distribution of the spray is still comparable. An example of this method of spray rate scale-up is as follows:

Scale-up (both units with the same type of spray gun)

	220 L bed	1,100 L bed
Number of spray guns	1	3
Number of nozzles per gun	3	3
Spray rate per gun (g/min)	1,000	1,000
Total spray rate (g/min)	1,000	to be determined

total spray rate = spray rate per gun (small unit) ×

$$\frac{\text{number of spray guns (large unit)}}{\text{number of spray guns (small unit)}}$$

$$= 1{,}000 \text{ g/min} \times \frac{3 \text{ spray guns}}{1 \text{ spray gun}} = 3{,}000 \text{g/min}$$

Scale-up (different types of spray guns)

	20 L bed	1,100 L bed
Number of spray guns	1	3
Number of nozzles per gun	1	3
Spray rate per spray gun (g/min)	50	to be determined
Total spray rate (g/min)	50	to be determined
Pattern diameter per nozzle (at 6 in. from the nozzle)	2.7	6.9
Pattern area per nozzle (in.^{2})	5.6	37.7

$$\text{ratio of pattern areas (per gun)} = \frac{\begin{array}{c}\text{\# nozzles per gun (large bed)}\\ \times \text{ pattern area per nozzle (large bed)}\end{array}}{\begin{array}{c}\text{\# nozzles per gun (small bed)}\\ \times \text{ pattern area per nozzle (small bed)}\end{array}}$$

$$= \frac{3 \times 37.7 \text{ in.}^2}{1 \times 5.6 \text{ in.}^2} = 20.2$$

Spray rate per gun = spray rate per gun (small bed) × ratio of pattern area (per gun) = 50 g/min × 20.2 = 1,010 g/min

Total spray rate = 1,010 g/min × 3 spray guns = 3,030 g/min

Process Temperatures

During scale up of the coating process, the product temperature is held constant. The spray rate is not generally scaled up in proportion to the increase in the fluidization air volume. Therefore, the standard approach to scale up is to adjust the inlet air temperature as required to maintain the same product temperature.

Potent Compounds

The pharmaceutical industry is developing more drugs that are extremely potent. These drugs require special considerations during manufacturing to protect operating personnel. The location of the air suspension system, the overall system, and the individual components must be reviewed for the manufacture of potent compounds.

The following items should be considered when a manufacturing facility for potent compounds is designed:

- The location of the facilities that will accommodate a system whereby operator involvement is minimized. This may involve several floors in the facility, where loading is performed on the highest floor, and discharge occurs on the lowest floor.

- Operator involvement must be minimized during the following stages of manufacturing:

  - Loading potent compounds into the air suspension system.

  - Unloading finished product.

  - Cleaning the equipment.

  - Cleaning and changing the filters in the inlet airstream, air suspension chamber, and the exhaust airstream.

- The air suspension system must be designed to provide minimum leakage at each seal and around the explosion vent. An air suspension system without an explosion vent is ideal for this application.

- The inlet and exhaust process ductwork for the air suspension system should be sloped, with the airflow velocity at a rate that will keep the particles airborne. This will prevent a buildup of compounds in the ductwork.

- The final inlet filter, all exhaust filters (including the dust collector), and the air suspension filters should be bag-in–bag-out filters.

- Redundant filters should be considered in the exhaust process airstream to ensure that product is not exhausted to the atmosphere.

- All filters and seals should be monitored for a plugged condition or a broken bag or seal.

Most manufacturers of air suspension systems will modify the air suspension unit by adding a contained loading chute and a contained discharge system to accommodate a particular application.

The loading may be by a gravity system from the upper floor or by the negative pressure created by an air suspension system. The discharge system can be a bottom discharge unit, where finished material is allowed to flow by means of gravity into a sealed container below the lower air suspension section.

During operation, personnel may be required to wear full body suits with breathing apparatus. Standard Operating Procedures (SOPs) should be written that will address the operating parameters for the process and personnel actions in the case of an accident. When the validation protocol is written for a system that handles potent compounds, issues associated with each product (e.g., controls and interlocks for loading, handling, bag in–bag out) must be reviewed and addressed in the protocol.

CURRENT DEVELOPMENTS/TRENDS

A current trend for coating equipment is the move toward improved control of the coating process. This is accomplished by adding mass flowmeters or mass balance systems that more accurately monitor and control the application of the coating solutions. The air handling systems incorporate humidification and dehumidification to control the drying rate consistently, regardless of seasonal variations. Control systems for coating systems are becoming more sophisticated and incorporate some level of automation.

Aqueous dispersions for the production of controlled-release films have been available since the early 1970s. However, it has only been fairly recent that these coatings have achieved widespread acceptance and use. Areas for further evolution of coating materials include increased solids level, improved controlled-release performance, and increased coating efficiency. One new technology, electrodeposition, creates a differential charge between the coating and the product substrate. The coating is attracted to uncoated surfaces on the substrate, resulting in the application of a uniform coating.

Another trend is toward a more economical coating process. This trend is being addressed in several ways:

- *Longer coating pans with less bed depth:* provides an increase in the total spray zone and results in a reduction in the process time.

- *High solids coating formulations:* reduces the overall quantity of coating formulation required to achieve the desired coating level.

- *Continuous coating pan systems:* used for high-volume production of noncritical coatings.

- *Vacuum or microwave systems:* increase drying efficiency for either perforated pan or fluid bed coatings.

REFERENCES

ASHRAE. 1996. Air cleaners for particulate contaminants. In *Heating, ventilation, and air-conditioning systems and equipment*. Atlanta, Ga., USA: American Society of Heating, Refrigerating, and Air-Conditioning Engineers, pp. 24.1–24.12.

Boucher, D. F. 1973. Fluid and particle mechanics. In *Chemical engineers' handbook*, 5th ed., edited by R. H. Perry and C. H. Chilton. New York: McGraw-Hill, Inc., pp. 5–54.

Campbell, R. J. 1997. Personal communication.

Chang, R. K. 1994. Use of a modified friabilator to assess logo abrasion during coating. *Pharm. Tech.* 18 (2):54–56.

Colorcon Inc. 1990. HT® Aluminum Lakes. Technology update, 6B-3 (10-90).

Crompton & Knowles. 1992. Chroma-Kote®DRI-KLEAR. Product Bulletin.

Dow Chemical. 1996. *Methocel cellulose ethers technical handbook.* Midland, Mich., USA: The Dow Chemical Co.

Flanders Filters. 1984. *HEPA filters and filter testing,* 3rd ed. Bulletin No. 581D. Washington, N.C., USA: Flanders Filters, Inc.

FMC Corp. 1986. *Aquateric® case history 2.* Philadelphia:FMC Corporation.

GAF. 1981. *Tableting with Plasdone®, Povidone USP.* New York: GAF Corporation.

Grain Processing Corp. 1995. *Pure-Cote® corn starches in aqueous film coating.* Technical Bulletin. Muscatine, IA, USA: Grain Processing Corporation.

Hercules. 1984. *Klucel, hydroxypropylcellulose, in aqueous film coating.* Wilmington, Del., USA: Hercules Incorporated Co.

Leaver, T. M., H. D. Shannon, and R. C. Rowe. 1985. A photometric analysis of tablet movement in a side-vented perforated drum (Accela-Cota). *J. Pharm. Pharmacol.* 37:17–21.

Morflex. *Pharmaceutical coating.* Bulletin 102-1. Greensboro, N.C., USA: Morflex.

Röhm GmbH. 1996a. Eudragit® enteric coatings pH control. *Basic info 1/E.* Darmstadt, Germany: Röhm GmbH.

Röhm GmbH. 1996b. *Sustained-release coatings with Eudragit RL/RS from aqueous dispersions.* Application Technology Sheet. Darmstadt, Germany: Röhm GmbH.

Sackett, G. L. 1997. Personal communication.

Seitz, J. A., P. M. Shashi, and J. L. Yeager. 1986. *The theory and practice of industrial pharmacy.* Philadelphia: Lea & Febiger, pp. 368–369.

Shinetsu Chemical. 1974. *Hydroxypropyl methylcellulose phthalate.* Technical Bulletin. Tokyo: Shinetsu Chemical Co.

APPENDIX: SOURCES FOR RAW MATERIALS AND EQUIPMENT

Included are alphabetized lists by trade name of commercial products referenced in this chapter. Separate lists are provided for coating materials and various types of equipment.

Coating Materials

Trade Name	Manufacturer	Address
Antifoam AF emulsion	Dow Corning Corporation	Midland, MI 48686
Aquacoat®	FMC Corporation	Philadelphia, PA 19103
Aquacoat®CPD	FMC Corporation	Philadelphia, PA 19103
Aquateric®	FMC Corporation	Philadelphia, PA 19103
Carbowax®	Union Carbide Corporation	Danbury, CT 06817
Chroma-Kote®	Crompton & Knowles ITC	Mahwah, NJ 07430
Chroma-Tone®	Crompton & Knowles ITC	Mahwah, NJ 07430
Citroflex®	Morflex, Inc.	Greensboro, NC 27403
Dri-Klear®	Crompton & Knowles ITC	Mahwah, NJ 07430
Eudragit®	Hüls America Inc.	Somerset, NJ 08873
Myvacet® 9-45	Eastman Fine Chemicals	Kingsport, TN 37662
Opadry I®	Colorcon	West Point, PA 19486
Opadry II®	Colorcon	West Point, PA 19486
Opaseal®	Colorcon	West Point, PA 19486
Opaspray®	Colorcon	West Point, PA 19486
PURE-COTE®	Grain Processing Corporation	Muscatine, IA 52761
Spectraspray®	Warner Jenkinson Company, Inc.	St. Louis, MO 63178
Surelease®	Colorcon	West Point, PA 19486

| Sureteric® | Colorcon | West Point, PA 19486 |
| Tween® 80 | ICI Pharmaceuticals Group | Wilmington, DE 19897 |

Air Suspension Coaters

Trade Name	Manufacturer	Address
Aeromatic®	Niro Inc. (Aeromatic-Fielder Div.) 21045	Columbia, MD
Flo-Coater®	Vector Corporation (Freund)	Marion, IA 52302
Fluid Air®	Fluid Air Inc.	Aurora, IL 60504
Glatt®	Glatt Air Techniques	Ramsey, NJ 07446
Kugelcoater®	Hüttlin (Distributed by Thomas Engineering Inc.)	Hoffman Estates, IL 60195

Batch Coating Pans

Trade Name	Manufacturer	Address
Accela-Cota®	Thomas Engineering Inc.	Hoffman Estates, IL 60195
Driacoater®	Driam USA, Inc.	Spartanburg, SC 29305
Fastcoat®	O'Hara Technologies Inc.	Ontario, CANADA M1M3T9
Hi-Coater®	Vector Corporation (Freund)	Marion, IA 52302
Procoater®	Glatt Air Techniques	Ramsey, NJ 07446

Continuous Coating Equipment

Trade Name	Manufacturer	Address
Continuous Coater®	Thomas Engineering Inc.	Hoffman Estates, IL 60195
Workhorse®	Coating Machinery Systems, Inc. (Vector Corp.)	Huxley, IA 50124

4

MICROENCAPSULATION

Robert E. Sparks
Irwin C. Jacobs
Norbert S. Mason

Particle and Coating Technologies, Inc.

In the last decade, the total cost of developing a new drug has risen dramatically, estimated by several authorities to be as much as 300–400 million dollars. Such a prohibitive cost makes it impossible for a middle-sized pharmaceutical company to carry through the complete drug development process. Alliances must be formed or limited rights must be purchased after the drug has been identified and approved. These are difficult paths, fraught with business problems, and they make it important to extend the usefulness of the drugs already being sold by a company.

One of the best methods of extending the market life of a drug is to modify its formulation in a useful fashion, such as extending the release of the drug, providing enteric protection to increase effectiveness, achieving a better masking of an off-taste, or targeting the drug to a particular part of the intestinal tract. Such changes will likely require that the drug be coated or microencapsulated, a major reason for the growth in research and development (R&D) in this field over the last 10 years.

This chapter will examine the major methods used for microencapsulation and the materials that are used in the process to modify the release of a drug or to protect it from its environment.

IMPORTANT CHARACTERISTICS OF THE CORE PARTICLE

Particle Size

Particles and Droplets above 800 μm

The particle size of the core has a great influence on both the process of microencapsulation that can be used and the control of drug release. For solid particles above roughly 800 μm, the pan-coating technique can be used, and the choice of this process becomes stronger as the particle size increases.

Fluid bed coating is another technique that also works well for large, solid core particles, even up to the size of tablets; however, the corners of coated tablets can be chipped during their motion in the fluid bed.

The organic-phase coacervation method also works very well for large, solid particles. However, this process is not appropriate for particles as large as tablets, especially where the contact of large particles with the soft walls of nearby particles during formation and drying would likely cause wall damage. This method is not known to be successful for large droplets.

Fluid drops up to approximately 3–4 mm in diameter can easily be coated using the annular jet methods of Southwest Research Institute and 3M company.

The spinning disk coating method of Particle and Coating Technologies works well for large solid particles; it is particularly useful for applying meltable coatings, even those with a high viscosity. The details of this method are discussed later.

Particles 100–800 μm

For particles above 100–150 μm, most processes will work well, but fluid bed coating should always be considered for solid particles, owing to its relatively low cost and good coating control. Cost considerations arise when the particles are very large, since fluidization of such large particles requires considerable energy. In addition, when the particles are below 200 μm, coating time can be long.

The organic coacervation process has few bounds in terms of particle diameter, and the process works very well for solid particles in this size range. Annular jet encapsulation can give good coatings for droplets as small as 250 μm in diameter under ideal conditions. The spinning disk coating technique is effective in this range, particularly when the coating can be applied as a melted material.

Particles 20–100 μm

The size range of 20 to 100 μm is important because the mean particle diameter must be in this size range to solve most taste-masking problems for liquid suspensions. Particles larger than this are usually judged to be gritty.

Only under ideal conditions can some of the latest fluid bed methods coat particles as small as 50 μm, for there is often some agglomeration of the fine particles, making coating times very long.

The organic coacervation method can coat these particles quite well. Spinning disk coating works well for particles as small as 40 μm; under ideal conditions, the particles can be as small as 20 μm, while still applying melted coatings.

Particles Smaller Than 20 μm

Only the coacervation method can coat particles smaller than 20 μm well, although particles can be made from suspensions of fine solids in a matrix material with mean diameters of 8 to 10 μm.

Size Distribution

The size distribution of the core particles has a significant effect on both the behavior of the coating process and the release characteristics of the particles after coating. For example, in fluid bed coating, a very narrow size distribution may lead to "slugging" of the bed, rather than the desired smooth fluidization. However, if the size distribution is very wide, many of the fine particles in the bed will be carried overhead into the particle recovery system. A good size distribution might be one in which the diameter varies by a factor of 3 from 5 percent to 95 percent of the cumulative volume as opposed to factors of approximately 7 to 9 for the typical wide distributions obtained from milling, spraying, or emulsification.

On the other hand, wide size distributions can be handled in the coacervation process. However, agglomeration of the finer particles or inclusion of the fine particles in the coating of the larger particles often occurs. These agglomerated particles produce thin spots over the outer portion of the agglomerate and, hence, give much less control of release than would be obtained from a narrower size distribution.

The size distribution in the annular jet method is formed as the jets break up, usually giving a narrow distribution of ±15 percent for the large drops, with a wider distribution for much smaller satellite drops. The latter usually have a mean diameter about one-fourth the diameter of the major drops, and generally consist of roughly 10–20 percent of the total mass of the droplets.

In spinning disk coating, the smallest core particles in a wide size distribution are agglomerated with the larger particles to form a single coated particle, giving the effect of thin spots. Thus, narrower distributions are ideal for this process, since it is possible to give essentially single particle coating for size distributions varying only two- to threefold.

Particle Shape

Solid particles that are roundish in shape, without protrusions or sharp corners, coat well in most solid-coating processes. However, when the core particles have been crushed or milled, they usually have sharp edges and corners. This can lead to thin spots in the coating, thus giving poorly controlled drug release or poor protection from the environment.

The process most able to handle these rougher particles is fluid bed coating, where many thin layers are put on the particle as it passes through a spray. Organic coacervation is also capable of putting reasonably uniform layers on such particles.

Solubility

The solubility of the core particle in a variety of liquids is perhaps the most important property for determining which process is best for coating. It is also important for determining the type of coatings that can be applied to give the desired release or protection. For example, if the core material has high water solubility in the coating solution, the coating may only be applied using the fluid bed process, where each thin, partial layer is dried as it is sprayed on. Hence, the solvent in the coating solution dries before it can dissolve the core particle.

For spinning disk coating, the particle must not have appreciable solubility in the melted coating at the temperature of application.

The most important effect of core solubility is in setting limits on the ability of any coating to confer protection or controlled release of the particle. If the core is highly water soluble, the coated particle cannot be placed in an aqueous liquid vehicle for storage, since the active ingredient in the core will be released into the liquid while the product is still on the shelf. Even if the coated particle is stored dry, there will be difficulty in obtaining taste masking (a short-term slowed release) of a small particle (say, 100 μm) in the mouth for a period of 5 to 10 min, since water will rapidly diffuse into the particle through the thin coating.

If there is a need to have the core released over a period of a few hours, a thick wall will be required, meaning that the particle diameter will have to be rather large (e.g., 800 μm).

The above considerations indicate that there are many controlled-release and protection problems that cannot be solved by a small coated particle. Situations continuously surface in which a 20 μm, water-soluble particle at 90 percent active ingredient must be coated, such that the coated particle can be immersed in water, but the active ingredient is protected from contact with water, for a period of 1 to 2 years. The solution to this problem requires that the coating have a diffusion coefficient for water that is many orders of magnitude lower than that of any known materials used as coatings.

Since the choice of coating materials is so important and requires the consideration of so many variables, this subject will be discussed first. A summary of the microencapsulation processes that are available will then be given.

SELECTION AND FORMULATION OF COATING MATERIALS

Approved Materials

Of the materials available to the formulation scientist, a very limited number of chemical entities are approved for use in pharmaceuticals. The number of polymeric substances is even more limited. The following discussion is not meant to provide a comprehensive listing of approved materials, because regulatory requirements vary from country to country; it will include the more common examples in each major classification.

Approved coating materials may be classified as water-soluble polymers, organic solvent–soluble polymers, latexes, and nonpolymeric substances. There is a large body of work from many laboratories describing the encapsulation of pharmaceuticals in widely varying materials, such as polyurethanes (Subhaga et al. 1995), glutaraldehyde cross-linked coacervates (Daniels and Mittermaier 1995), chitosan (Akbuga and Bergisadi 1995), polyhydroxyethyl methacrylate copolymers, and poly n-isopropyl acrylamide. While these studies are valuable from the point of view of methods development and prove the need for the development of more useful polymer systems, these coatings are not presently acceptable and will not be covered in detail in this review. Table 4.1 is a quick reference with some abbreviations that are explained in the text.

Table 4.1. Approved Coating Materials

Water Soluble Polymers		
Polysaccharides		**Nonpolysaccharides**
Cellulosics	**Noncellulosics**	
NaCMC	guar gum	gelatin
MC	carrageenan	PVA
HPC	karaya gum	PAA & XL'ed*
HPMC	gum arabic	PEG
CAP	sodium alginate	PEG co-PPG
HPMCAP	chitosan	PVP & XL'ed*
CAT		

Organic Soluble Polymers	Latexes, Dispersions or Pseudolatexes	Nonpolymers
ethylcellulose	methacrylic acid co-ethyl acrylate	hydrogenated vegetable oil
CAB	ethyl acrylate co-methyl methacrylate	vegetable waxes
shellac	ethylcellulose + plasticizers	linear fatty acids
methacrylates	CAP	sugars
PLA and PLA/PG		
PEG esters		
zein		sorbitan esters
CAP, HPMCP		
CAT, HPMCAS		
polyvinyl acetate phthalate		

*Cross-linked polymers technically are not water soluble but are highly swollen.

Water-Soluble Polymers

Water-soluble polymers can be further subdivided into polysaccharides and nonpolysaccharides, with the polysaccharides further divided into cellulosics and noncellulosics. The cellulosics useful in microencapsulation include sodium carboxymethyl cellulose (NaCMC), methylcellulose (MC), hydroxypropyl cellulose (HPC), hydroxypropyl methylcellulose (HPMC), cellulose acetate phthalate (CAP), hydroxypropyl methylcellulose acetate phthalate (HPMCAP), and cellulose acetate trimellitate (CAT). The last three materials are only soluble in aqueous ammonia.

Noncellulosic polysaccarides include guar gum (galactose and mannose units), carrageenan (sulfated linear polysaccharides of D-galactose and 3,6 anhydro-D-galactose), karaya gum (partially acetylated complex polysaccharide), gum arabic (complex polysaccharide containing L-arabinose, L-rhamnose, D-galactose, and salts of D-glucuronic acid), starches (soluble and cross-linked), sodium alginate (sodium salt of copolymer of mannuronic acid and guluronic acid), and chitosan (de-acetylated chitin).

The nonpolysaccharide water-soluble polymers include gelatin (protein), polyvinyl alcohol (PVA), polyacrylic acid (PAA) (linear and lightly cross-linked), polyethylene glycol (PEG), polyethylene glycol copropylene glycol (PPG), and polyvinylpyrrolidone (PVP) (linear and cross-linked). The cross-linked polymers are not water soluble but are very swollen networks of very high molecular weight molecules.

The use of these water-soluble polymers is quite varied, depending on their ability to act as granulation binders, film modifiers, disintegrants, bioadhesives, or film-forming agents. High isoelectric point gelatins (from acid treated pig skin), and the more recently studied chitosan, are cationic polymers. Thus, when used in conjunction with anionic polymers such as gum arabic, sodium alginate (Arneodo et al. 1986), or type B gelatin (Remunan-Lopez et al. 1995), they form swollen ionic complexes known as coacervates for encapsulation. The coacervate coatings are normally cross-linked either thermally or with difunctional reagents such as glutaraldehyde to insolubilize and toughen them. The capsules can then be filtered and dried or the water removed by lyophilization. However, the chemical cross-linking treatment may render these microcapsules unacceptable under current standards for pharmaceuticals. This may limit the utility of coacervates for pharmaceutical production.

Many of the cellulosics, such as MC, HPC, and HPMC, are available in varying molecular weights, which can affect their behavior as disintegrants or granulation aids. The higher molecular weight versions are preferred as disintegrants. The highly substituted, low-viscosity grades are preferred for film coating or in sealing cores prior to sugar coating. Films formed from these polymers are flexible with good adhesion, are good barriers to oily substances, and can reduce friability in particulate systems. HPMC has more recently been used in matrix formulations to give controlled-release properties (Gao et al. 1996). HPC coatings on vaccine particles have been found to protect the antigen from solvent degradation during a multiple emulsion encapsulation process (Lee et al. 1996).

Organic Solvent–Soluble Polymers

Acceptable organic solvent–soluble polymers are far fewer in number, and it seems that there is little intent in the industry to expand the number further. Several organic solvent–soluble polymers are being prepared as latexes or pseudolatexes to avoid the use of organic solvents in encapsulation processes. The organic solvent–soluble polymers include ethylcellulose; cellulose acetate butyrate (CAB) (Arnaud et al. 1995); shellac (soluble in ethanol or in aqueous base); meth-acrylate and acrylate copolymers; poly L-lactic acid; poly D,L-lactic acid; poly D,L-lactic co-glycolic acid; zein; and several esters of dicarboxylic acids, including CAP, hydroxypropyl methylcellulose phthalate (HPMCP), CAT, hydroxypropyl methyl cellulose acetate succinate (HPMCAS), and polyvinyl acetate phthalate.

Ethylcellulose has been used in a variety of processes for coating or encapsulating small particles. The polymer is available as a solid in a variety of molecular weights and several levels of ethoxyl content. This can translate into adjustments in barrier performance or other process requirements. Higher molecular weights (higher solution viscosities) tend to yield stronger, tougher films and, thus, improved barrier performance. Lower levels of ethoxyl content (45 to 46.5 percent) tend to yield tougher films, although they have somewhat poorer solvent solubility. Ethoxyl contents in the 48 to 49.5 percent range tend to produce slightly lower melting and softening point ranges and have better solvent solubility. The polymer is readily soluble in a wide variety of solvents, including ethyl alcohol, methylene chloride, acetone, isopropyl alcohol, toluene, and ethyl acetate. Often, the barrier properties of the films applied from solvents are superior to those obtained with aqueous dispersions at equivalent amounts of polymer and molecular weight.

The pseudolatexes tend to contain lower molecular weight polymers than are available for organic dissolution. The polymer can be applied using fluid bed coating techniques or a variety of coacervation processes. Solvent-free coating mixtures of ethylcellulose have been prepared by blending it with an excess of meltable plasticizer, such as stearic acid or stearyl alcohol (Scott et al. 1964). This melted mixture can then be applied to small drug particles in either fluid bed or spinning disk coating processes.

It should be noted that the films applied from differing solvents or solvent mixtures have different tensile strengths and percent elongation at break; they will give different release characteristics when applied as coatings on small particles. An optimal polymer solution will yield the maximum unwinding of polymer chains and produce films with optimal mechanical properties (Banker 1966).

There are a number of methacrylate copolymers available for oral dosage forms. These include the water-swellable methacrylate ester copolymer with trimethyl ammonioethyl methacrylate chloride, the gastrosoluble film of the copolymer of a methacrylate ester and dimethylaminoethylmethacrylate, and the neutral methyl methacrylate co-methacrylic acid. These are commercially available under the trademark Eudragit® and are available either as solids or as lacquer solutions in volatile solvents.

A wide variety of biologically active materials have been successfully encapsulated in poly D,L lactic acid or poly D,L lactic co-glycolic acid for sustained release (e.g., Lupron Depot [TAP Pharmaceuticals, Deerfield, Illinois]). Various preparation methods have been described, but solid-in-oil-in-water methods or water-in-oil-in-water (w/o/w) double emulsion methods (Jeffery et al. 1993) are the most common, with evaporation of a methylene chloride solution of the polymer providing formation of the microcapsules. Polymers such as PVP or PVA are often used to stabilize the dispersions along with surfactants such as phosphatidyl choline or sorbitan esters. Low entrapment efficiencies are often seen, although recently there have been developments to improve efficiencies substantially (Uchida et al. 1996). Alternatively, spray drying of the solid-in-organic dispersion or water-in-organic emulsion can result in successful encapsulations under more controlled conditions of temperature, solvent selection, and polymer concentration. The release rates of the active substance can be influenced by a variety of factors, including polymer molecular weight and comonomer ratios. Since the principal mode of release is the slow degradation of polymer by ester hydrolysis, the initial release rate is based on the slow diffusion of water into the amorphous zones of the polymer, followed by random hydrolysis and eventual failure of the polymer. Hydrolysis and release are affected by molecular weight, the crystallinity and the hydrophilicity of the polymer, environmental pH, and temperature.

Zein, a prolamine from corn gluten, is soluble in isopropanol and 70/30 acetone/water, but it is insoluble in neutral aqueous solutions. However, zein has been found to swell in aqueous acid; thus, it can be used as an undercoat or extended release agent with reverse enteric properties.

Shellac, a refined resinous excretion of the insect *Laccifer lacca*, has not found wide acceptance in encapsulation due to its variability in barrier performance based on methods of purification and batch consistency. Changes in the release profile after extended storage have also been noted (Müller and Yunis-Specht 1994). Shellac is soluble in alcohol, aqueous ammonia, or alcoholic amine mixtures. Its primary use is in tablet sealing or enteric coating.

There are a variety of phthalate esters applicable in solvent-based coating processes, including CAP, CAT, and HPMCAP. The latter two have the added advantage of being soluble in ethanol/water mixtures (85:15), while all of the phthalate esters are soluble in a variety of solvent mixtures including acetone:ethanol, methylene chloride, and ethyl acetate:ethanol. HPMCAP is a more flexible polymer than CAP or CAT and may not require as much plasticizer.

With the high cost of solvents, solvent recovery systems, air quality controls, and the handling problems associated with explosive and toxic solvents, pharmaceutical manufacturers are moving in the direction of water-based systems. However, there are drawbacks to aqueous-based systems, including longer processing times (higher heat of vaporization), the need to control film formation from latexes or aqueous dispersions, and other problems involving the coating of small particles or coating highly water-soluble materials with the aqueous systems.

Latexes and Dispersions

Aqueous polymeric dispersions can be formed in a variety of ways. A latex is prepared by the polymerization of an emulsion of oil-soluble monomers in water. Examples include the copolymers methacrylic acid co-ethyl acrylate and ethyl acrylate co-methyl methacrylate, available as Eudragit® L30D and Eudragit® NE30D, respectively. Pseudolatexes are dispersions prepared from solutions of the polymer by one of several emulsion or emulsion/solvent evaporation processes. Ethylcellulose pseudolatex is available as Aquacoat® ECD (contains cetyl alcohol and sodium dodecyl sulfate) and Surelease® (contains dibutyl sebacate, oleic acid, ammonia, and fumed silica). CAP is available as a pseudolatex that has been spray dried and contains Pluronic F-68 (a PEG/PPG copolymer), Myvacet® 9-40 (an acetylated glyceride), and Tween® 80 (a PEG sorbitan mono-oleate). A 30 percent solids aqueous dispersion of CAP is also available as Aquacoat® CPD. Polyvinyl acetate phthalate is available as a micronized polymer (Sureteric®), which also contains pigments and plasticizer.

Nonpolymeric Coating Substances

Nonpolymeric substances or insoluble polymeric additives can be included as all or part of a controlled release formulation for preparing small particles. Molten materials, such as hydrogenated vegetable oils, vegetable waxes, stearic or behenic acid, stearyl alcohol, tristearin, monoglycerides, hydrogenated castor oil, or microcrystalline waxes, can be used in a variety of processes, including pan coating, fluid bed coating, spray congealing, or spinning disk coating, to form either matrix particles or capsules with a single core. The use of stable, well-characterized, and well-controlled supply sources are critical when using these "naturally derived" substances. However, they can provide uniform performance at minimal cost if the process parameters are carefully controlled. Using these meltable materials for encapsulation can provide inexpensive means of delivery for new dosage forms, because there is no solvent evaporation, and only the heat of fusion needs to be removed in the process. The performance of these materials often depends on understanding their crystalline phases during solidification and properties, such as hydrophobicity and permeability. This is particularly critical when using triglyceride coatings, where transformations between crystal forms can take place in storage and change the barrier performance (Siekmann and Westesen 1994). The addition of compatible polymers or particles can often greatly alter the permeability of these coatings to give a desired release profile.

Sugars, such as lactose or glucose, can be used to modify the permeability of films in aqueous film coating or suspensions in molten coatings. Release rates can readily be increased with relatively small amounts (<10 percent). PEG mono- or diesters, PEG castor oil, or sorbitan esters can be similarly used to modify the release profiles of hydrophobic barriers. They usually increase the rate of release but may also improve the total release of the active if there is a plateau of release at values well below 100 percent. Mannitol and trehalose have been found to increase the stability of protein formulations against organic solvent–induced denaturation during microencapsulation in biodegradable polymers (Cleland and Jones 1995).

Using small amounts of disintegrants, such as cross-linked NaCMC, cross-linked PAA, cross-linked PVP, or sodium starch glycolate, can provide desirable effects in the release profiles of microcapsules. If the disintegrant is contained within a matrix formulation or the core of a core/shell formulation, the disintegrant can provide a short delay or a lower release rate, while there is water imbibition and swelling of the polymer. This is followed by an accelerated release rate as the particle begins to crack open and increased surface is exposed for release. This type of behavior has been studied in hydrogenated trigylceride matrixes exhibiting a type of erosion release and near zero-order rates (Morello-Mathias 1995). If water-swelling excipients are included in a hydrophobic barrier coating of a microcapsule, the coating can still provide a barrier to atmospheric moisture or a physical barrier to prevent ingredient interactions, yet accelerate release of the active ingredient by cracking off the coating. Polymers that swell differentially in response to ionic strength, pH, or other stimuli may provide avenues for triggered response formulations.

Criteria for Selecting Formulation Ingredients

The following criteria may be used to help select candidate materials and processes for use in encapsulation: the purpose of the microencapsulation (taste masking, enteric or reverse enteric release, time release, dosage form design constraints, etc.), the barrier properties of various coating or matrix materials, formulation effects based on desired properties in the product, and requirements based on needs to improve processing.

Purpose of Microencapsulation

Taste Masking. Taste masking of small particulate drugs has a number of conceptual difficulties. Mouth-feel considerations dictate that the mean diameter of the coated particles be under 120 μm. For hard coatings, the mean diameter of the coated particles may have to be below 80 μm. This limits both the type of process that can be used and the materials that are suitable. Fluid bed coating becomes considerably more difficult as core particle sizes below 100 μm are considered, since particle agglomeration and the reduction in the rate of coating become significant operational problems. However, recent improvements in the methodology and equipment have lowered the particle diameter for good fluid bed coating.

In taste masking, the release of the active ingredient should be prevented for 5 to 10 min to be sure the coated drug particles are beyond the palate. However, there should still be 100 percent bioavailability. This may require some trigger release mechanism, such as the use of delayed disintegrants, or another triggering mechanism based on the change of environment from the stomach to the upper intestine. If the taste-masked formulation is to be dosed as an aqueous suspension, then the coating must remain intact in water, which precludes many possible formulations. Spinning disk coating and spray congealed matrix preparations are finding wider application in this area.

Enteric and Reverse Enteric Release. Enteric release is based on materials that are resistant to acidic gastric fluids and release in the neutral pH of the upper intestine. The methacrylate copolymers and the phthalate esters have been widely used for their enteric protection. Reverse enteric coatings are hydrophobic and neutral above approximately pH 6.5, but they become charged and more hydrophilic at gastric pH (< pH 3). Styrene co-vinylpyridine has shown this type behavior but has not been approved for human use. Zein, although neutral, has shown some acceleration of release at gastric pH. Cationic acrylic polymers are more soluble at low pH than neutral conditions, although they are swellable even under neutral conditions and may offer some advantages for taste-masking formulations.

Timed Release. Timed-release coatings are used to deliver a particular medication over a defined interval, whether it be first-order or zero-order kinetic release, or provide a defined delay—the goal in an encapsulated booster dose of vaccine. If extended-release regimes are used, consideration of complete release within the gastrointestinal transit time must be taken into account.

Dosage Form. The desired dosage form of a drug can limit one's choices of encapsulation materials or methods. The coating of small particles for subsequent tableting requires that the coating be significantly robust to withstand the rigors of the tableting process. Often, the coated particles must be less than 200 μm in diameter to avoid significant coating fracture during tableting, although changes in the tablet formulation can permit the use of larger coated particles. Toughening the wall requires the use of plasticizers in the coatings. Matrix particles can readily be damaged during tableting and may require tablet excipients such as PEG or microcrystalline cellulose (Sakr and Oyola 1986). Microcapsules for injection have many more restrictions based on particle size and the type of coating that can be used.

Barrier Performance of Coatings

Coating materials are often chosen for their performance as barriers to the diffusion of water into the particle, and the dissolution of the active followed by diffusion of water and solute from the particle. However, factors such as coating strength, thickness, and toughness are not taken into account. Resistance to osmotic rupture can be a key factor in controlling the rate of release of an active ingredient. Generally speaking, diffusional release through a barrier coating decreases with increasing coating thickness. Also, the rate of water diffusion through a barrier can be decreased by reducing the interstitial spacing between molecules of the coating by using higher molecular weight coatings (Rowe 1986), higher density coatings (closer molecular packing), or particulate fillers.

Economic factors play a role in choosing materials and barrier performance. The cost of the coating material is not as significant as the labor and processing costs; however, a balance will usually be required between the cost of the coating formulation and the thickness of the coating required.

Barrier performance requirements based on changes in pH, temperature, ionic strength, or other environmental factors can dictate material selection. Enteric and reverse enteric polymers were mentioned above. Several different types of hydrogels based on copolymers of 2-hydroxyethyl methacrylate co-methacrylic acid (Brannon-Peppas and Peppas 1990) and n-isopropyl acrylamide co-methacrylic acid (Baker and Siegel 1995) exhibit differential swelling and, thus, barrier properties in response to pH changes. The latter polymer also exhibits differential swelling in response to temperature. The switching temperature of the n-isopropyl acrylamide copolymer gels have been studied to achieve regulation in response to small temperature changes around body temperature (Yoshida et al. 1994). Some polymers respond to ionic strength or the concentration of particular cations. Pectin, alginate, gellan gum (tetrasaccharide repeat units), carrageenan, and polyacrylates can be useful as constituents of barrier coatings that need to swell or shrink in the absence or presence of particular ions.

Formulation Effects on Desired Properties in the Product

There is a large body of information showing the effects of formulations on release rates. Many of these effects, although somewhat intuitive, have been confirmed in testing and can be summarized as follows.

Surface Active Agents. The addition of surface active agents to hydrophobic barrier coatings will increase the rate of release of a water-soluble active ingredient (Hamid and Becker 1970).

Water-Soluble Polymers. The addition of small amounts (1–5 percent) of water-soluble polymers to hydrophobic barriers can increase the rate of release of water-soluble active agents (Donbrow and Friedman 1974). The addition of 10 to 15 percent water-soluble polymers, such as gums, may not affect the water vapor transmittance of a hydrophobic barrier, but it may facilitate the disintegration of the barrier when the coated particles are submerged in an aqueous environment. Such coatings may be used to isolate incompatible ingredients in a tablet formulation and may be useful in programming a triggered response to a change in pH, ionic strength (Brannon-Peppas and Peppas 1991), or ion concentration. The use of small amounts of cross-linked polymer excipients, such as cross-linked CMC or PAA can provide a similar or even more accelerated disintegration of coating when placed in an aqueous environment.

Plasticizers. Plasticizers are often required as additives to polymeric coatings, especially in fluid bed coating. This requirement is to ensure the elasticity of the films and to reduce the tendency of the cores to stick together during processing. The plasticizers are especially important when latexes or pseudolatexes are used, since they promote the fusion of the coating into a more uniform barrier by lowering the glass transition temperature (T_g) during processing. The addition of a plasticizer also decreases the barrier properties of a coating, since rotational freedom of the coating molecules is increased as the T_g is decreased. However, in many cases, observed release rates are lower because there is a decreased tendency for osmotic rupture with the increased flexibility in the barrier films. Films with added plasticizer are more flexible and are not damaged as easily in tableting operations.

Each plasticizer has a different resistance to water and may change the overall water permeability by varying degrees. Effective plasticizers tend to have a structure similar to the polymers to be plasticized. Thus, hydrophilic cellulose ethers are best plasticized by polyols. Polymers with strong intermolecular cohesive forces may require larger amounts of plasticizer, up to 20 to 40 percent (Deasy 1984). The more hydrophobic plasticizers, such as dibutyl sebacate, are not water soluble and require small amounts of emulsifiers in aqueous systems.

Particulate Fillers. Many particulate fillers, such as talc or magnesium stearate, can improve the sealing effect of the film-coating process (Chatfield 1962). In encapsulation processes, bentonite can improve the barrier by filling in the interstitial spacing of the coating molecules and may provide a slight time-delayed acceleration of release as the bentonite absorbs water and begins to crack the coating.

Environmental Considerations. Environmental considerations can often dictate the use of one solvent system over another or force the use of pseudolatexes or dispersions. Methylene chloride is the solvent of choice for polylactic acid and its glycolide copolymers. The residual solvent can be reduced to very low levels. Cyclohexane is required in significant quantities in ethylcellulose coacervation processing (Morse and Hammes 1975), thus requiring careful economic and environmental considerations for each application. The barrier properties of hydrophobic polymer coatings can vary greatly with the solvent used in the processing. Improved film formation is favored by the solvent that promotes better solubility (uncoiling) of the polymer.

Flow Aids. Flow aids are often used in encapsulation processes, where the product may go through a tacky phase before the product is fully dry and free flowing. In coacervation processing, flow aids are added after the filtration step. In spray drying, the drying polymer goes through a tacky phase and can impact another particle and produce agglomerated product. To prevent this, silica can be added to the drying air. Talc is often added to latex formulations in fluid bed coating of small particles to help prevent agglomeration.

Process Improvement

The matching of coating materials to a coating process will be largely controlled by the limitations of each process: the viscosity of the molten or dissolved coating, the requirements for additives and processing aids, the suitability of solvents, and the ability of the process to permit the use of formulated coating mixtures, such as solid-laden coatings.

EQUIPMENT AND COATING PROCEDURES

Given the need for microencapsulation, the goal is to select a process that can meet the product specifications at a minimum cost.

The factors that may be important in cost reduction are raw material cost, equipment availability, familiarity of the staff with a particular process, ease of regulatory approval, and ease of adapting the process to the product specifications (e.g., release rates and bioavailability). The size as well as the size distribution of the required particles will influence the decision. If submicrometer particles are desired, then a process in which the continuous phase is a liquid is preferred, because emulsification is much less expensive than atomization. In other cases, it may be practical to combine emulsification with atomization, as done in encapsulating flavors or fragrances in matrix particles.

Continuous Phase Is Air or a Gas

Particle Size Is Determined by Spraying

Spray Chilling. In spray chilling, a liquid is atomized into cold air or a gas to solidify the droplets. The liquid droplet may contain solids or an emulsified second liquid phase that then becomes encapsulated as the droplet solidifies. The amount of encapsulated material may be 40 percent or higher, although it is more typically 20–25 percent. The process vessel is often a spray dryer without the heater activated. Product collection can be in the chamber of the vessel itself, in a cyclone, or both. In the pharmaceutical industry, a filter bag is often used to remove the fines from air being exhausted. The dust that is collected is usually discarded. Melts and compositions that gel on cooling are used most often as the continuous phase. The liquid may be atomized with a swirl type pressure nozzle, a rotating disk, or a two-fluid nozzle. In general, for more uniform sprays at high capacities, rotating disks are more economical. Two-fluid nozzles are preferred for particle sizes below 50 μm and for low production rates. However, the cost of compressed air is usually significant. The factors that determine the choice of the atomizer are discussed by Lefebvre (1989), Marshall (1954), and Masters (1979). The feed lines usually need to be heated to a controlled temperature. This is sometimes necessary for the lines carrying the atomizing air or atomizing gas. Normally, ambient air is sufficiently cold for solidifying the droplets, depending on the melting point of the coating, although higher production rates may require additional chilling of the air. The higher the melting point and the smaller the droplets, the more rapid the solidification of the drops. Droplets as large as 700 μm can be solidified in ambient air when coatings melt in the range of 110 to 120°C. With melting points in the range of 60°C, particles of 100 μm can be solidified in equipment only a few feet high.

Spray Drying. Spray dryers are widely used in pharmaceutical and biochemical processing, but such processing often presents problems of a regulatory nature and may require special controls that are not encountered with other uses of spray dryers (Masters 1991). Common requirements are that the surface finish on the #360 stainless steel be electrically polished, that aseptic conditions be maintained, and that drying air and atomizing air be filtered with HEPA (high efficiency particulate air) filters. Also, the ease of cleaning is given high priority.

In spray drying, a liquid is atomized in a vessel in which the droplets contact hot air or gas. Sufficient heat is transferred to cause evaporation of a portion of the liquid present so that the partially dried particles do not adhere to each other or to the walls of the vessel. If the droplets contain solids or a nonvolatile liquid phase, as the volatile solvent evaporates, the dispersed phase becomes encapsulated in the solidifying continuous phase.

Figure 4.1 is a schematic of an encapsulation spray drying process. The upper limit of the dispersed phase is between 40 and

Figure 4.1. The pharmaceutical drying encapsulation process.

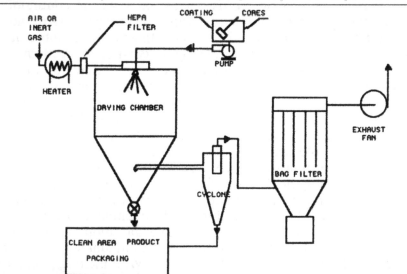

50 percent. If each particle contains several particles of the active core, a matrix particle results; if it contains a single core, it is a microcapsule. Collection of the product is similar to that in spray chilling. Spray dryers are usually run with concurrent flow of air and particles to give minimal overheating of the particle. This is important if the contents are heat sensitive or somewhat volatile. The temperature of the droplet at any point in the dryer corresponds to the wet bulb temperature of the gas phase surrounding the droplet. However, the concurrently dried particles are likely to be more porous than if the dryer were run countercurrently. In the countercurrent case, the wet particles contact air of a somewhat lower temperature, causing slower evaporation of water and the formation of more compact films. The dry particles near the bottom contact the hottest air; thus, countercurrent processing requires more thermal stability in the particle.

The dimensions of the spray dryer determine the maximum size of droplets that can be dried. For example, if the droplets are too large, they will impinge on the wall and adhere to it. A 1 m diameter laboratory dryer is usually adequate for 50 to 60 μm droplets from a two-fluid nozzle; for 100 μm droplets, a 2 to 2.5 m diameter dryer is required. There is no physical limit to the dimension of a spray dryer or the size of particles that can be dried. Rather the size limit is set by economics and the availability of space. If particles larger than 250 μm are needed, it is usually more economical to use an agglomeration step after first producing smaller particles. For instance, an auxiliary system, such as a fluid bed granulator, can be connected to the spray dryer and designed to become an integral part of an overall system.

Atomizing nozzles are usually mounted to spray downward. However, it is also possible to mount the nozzle to spray upward (fountain spray), which permits somewhat larger drops to be dried because the residence time of the droplet is greater. The residence time is on the order of a few seconds. A dryer is generally rated according to the amount of water that can be evaporated per hour. Disregarding tabletop dryers, the smallest dryers are rated for 10 lb/h. In spray drying, the common heat sources are natural gas or electricity. Natural gas is generally more economical, but electricity is more versatile since special systems such as inert gas drying and closed systems using special solvents can be implemented more economically.

The pharmaceutical applications of spray drying are numerous, including for antibiotics, digitalis, hormones, gamma globulin, and blood fractions. Encapsulation occurs when a dispersed phase is present.

Annular Jet Method. Microcapsules of a uniform size distribution can be made using concentric tubes. The core is pumped through the central tube, while the coating material is pumped through the outer annular space. The velocities of the two fluids are carefully matched at the interface. The concentric jet breaks up into droplets due to Rayleigh instability caused either by spontaneous perturbations on the surface or by deliberately induced pulsations (Samuels and Sparks 1973). According to theory (Rayleigh 1879), the diameter of the droplet is 1.9 times the inside diameter of the tube, although values as high as 2.1 have been reported (Merrington and Richardson 1947). The tubes can be stationary or rotated so that centrifugal force can amplify the effect of the fluid pressure (centrifugal extrusion) (Summerville 1962). This process has been brought to a practical level by Southwest Research Institute (San Antonio, Texas). Heads with up to 16 nozzles are available. The process can form microcapsules from 400 to 2,000 μm, which are within 10 percent of the mean diameter up to a production rate of 22.5 kg per nozzle per hour.

If the coating has been melted, the droplets may simply be solidified by cooling as they fall through air. If subsequent treatment is needed, as in the gelation of a sodium alginate wall in calcium chloride, the drops can be caught in a hardening bath. These techniques have been used to encapsulate water in wax and oils in gels such as alginate.

Coating a Preexisting Particle

Fluidized Bed Coating. Figure 4.2 shows schematics of three common designs. This method is widely practiced in the pharmaceutical industry for particles in the range of 100 μm up to the size of tablets (Deasy 1984). The particles to be coated are "fluidized" by means of air or a gas flowing from below. Two-fluid nozzles are used almost exclusively to atomize the coating. The production of a uniform coating requires good circulation of the particles beyond the nozzle. There are three common positions for the nozzle: on the bottom of the bed spraying upward, on the top of the bed spraying downward, and tangentially (spraying from the side). When spraying solutions and latexes, bottom and tangential spraying often gives better efficiencies and better distribution of the spray because the spray has more opportunity to contact the particles. Spraying efficiencies can approach 100 percent. Droplets sprayed from the top of the bed downward may be carried away from the bed by the fluidizing air or can dry before reaching the particles to be coated; however, top spraying is useful in applying melt coatings.

Figure 4.2. Schematics of fluidized bed coating: (a) top spray coater. (b) Wurster (bottom spray) coater. (c) rotor (tangential spray) coater. Reproduced courtesy of Glatt Air Techniques.

(a)

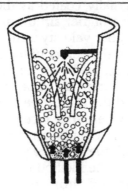

(b)

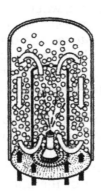

(c)

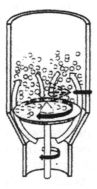

The Wurster coater uses a draft tube suspended above the spray nozzle at the bottom of the apparatus. The particles flow inward, under the edge of the draft tube, from the annular region outside the draft tube. They are partially coated as they move upward in the draft tube, propelled by a higher airflow as well as by air from the two-fluid nozzle. The higher velocity in the draft tube is created by having larger openings in the center of the air distribution plate. After moving upward, drying in the air, the particles disengage in an expanded section above, then fall back into the annular zone between the outer wall and the draft tube. The particles then make a return trip to the nozzle where they receive more coating. This gives good control over particle circulation, thus, producing more uniform coatings. In general, the Wurster process gives less erosion of particles, which decreases the likelihood of agglomeration.

Coatings can be applied from water, a solvent, a latex or pseudolatex, or a melt. Dispersions and latexes can be applied with solid contents up to 30 percent, whereas solutions of high polymers need to be more dilute (often less than 10 percent) because the solution viscosity is higher. Latexes and dispersions require plasticizers to allow the dispersed phase to fuse as it forms the coat. If the granules being coated are later to be compressed into tablets, plasticizer may also be added to decrease breakage of the wall during tableting.

Polymers need to be above their T_g so that they can fuse. Plasticizers function by lowering the T_g. Coatings from solvents are more expensive because of the cost of the solvent and the regulatory and environmental problems associated with them. However, solvents are often necessary because of the improvements obtained in the barrier properties. Glatt Air Techniques (Ramsey, New Jersey) makes Wurster coaters that have a product volume up to nearly 800 L, a nominal diameter of 1.8 m, and a height of 6.1 m; they can use 10,000 m^3 of air per hour.

Pan Coating. Pan coating (Porter and Haluska 1991) is an older process that has been used extensively to coat tablets and candies. It is similar to fluidized bed coating; particles are preformed, and the coating is applied in the form of a spray. Air is drawn through a bed of particles that is rolled by the rotation of the pan. The pan generally uses less air than the fluidized bed. Air can be supplied through sword-shaped tubes immersed in the bed of particles and then drawn off by an air handling system. This is used either to remove solvents that are part of the coating formulation or to eliminate the heat of melting of melt coats. The coating pan has seen considerable development, and there are a variety of models available.

Spinning Disk Coating. When used with melt coatings, this process has the advantage of not requiring solvents or water and of being able to coat particles individually. It is a flexible process because coatings with viscosities upward of 1000 cp (or more viscous for larger particles) can be applied. It is a high-volume, low-cost process. Core particles in a preferably narrow size range are mixed with a coating that consists of polymers such as ethylcellulose or ethylene vinyl acetate copolymers dissolved in a hard fat or wax composition. The slurry is pumped to a rotating disc that separates the coated particles from each other and from excess coating. Figure 4.3 illustrates the process involved. The coating is usually solidified by cooling in a vessel similar to a spray dryer. It is similar to spray chilling except that cores of a known size are supplied separately, and their diameter defines the size of the product. The rotating disc is operated at a speed designed to produce coated product and excess coating particles that are smaller than the product. The excess coating is then easily removed by sieving and recycled (Sparks and Mason 1987) if allowable.

Continuous Phase Is a Liquid

When the continuous phase is a liquid, a coating is deposited on a dispersed droplet. The size of the droplets of the dispersed phase generally determines the size of the final microcapsule (except for matrix particles containing fine dispersed droplets). These droplets may be generated by shear, using a turbine in a baffled vessel (Shinar and Church 1960; Shinar 1961), a high-shear rotor/stator mixer, static mixers in a pipe (Maa and Hsu, 1996), a colloid mill, a homogenizer, or by pumping the liquid through appropriately sized tubing or openings.

Aqueous Continuous Phase

Aqueous methods are best suited for the encapsulation of materials that have minimal water solubility, such as oils, since they must form the dispersed phase in the process. Examples in the pharmaceutical industry are few: taste-masked cod liver oil, oily organic dyes for imaging, and oil-soluble vitamins. The reason these processes are not often practiced in the pharmaceutical industry is that most medicines are solids and are often more soluble in water than in oils.

Simple and Complex Coacervation. The polymer is dissolved in the aqueous medium and the nonpolar or oily phase is dispersed by means of mechanical agitation. A change is made in the medium

Figure 4.3. Spinning disk process schematic. Reproduced courtesy of Glatt Air Techniques.

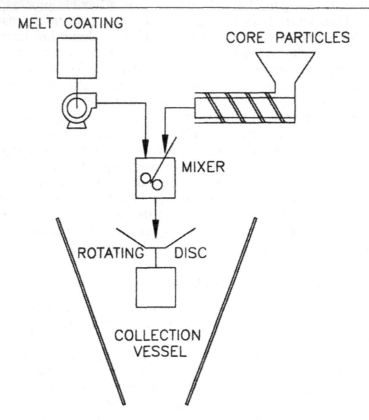

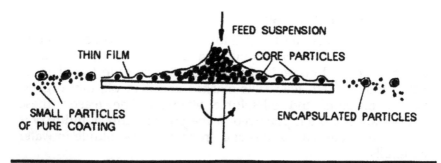

(pH, ionic strength, temperature, etc.) to induce the formation of a second polymer-rich phase that contains the eventual wall material. In the case of a protein such as gelatin with simple coacervation, the polymer-rich phase can be induced by adding ethanol, sodium sulfate, or other solutes that are very water soluble and can compete for the water (Khalil et al. 1968) of solvation.

Complex coacervation is achieved, as indicated by Figure 4.4 that describes the encapsulation of cod liver oil, by lowering the pH of a solution containing two polymers, one of which gains positive charges as a result of the change (e.g., pigskin gelatin of high isoelectric point), while the other polymer has only negative charges [e.g., gum arabic] Bungenberg de Jong 1949). The ensuing

Figure 4.4. Complex coacervation process.

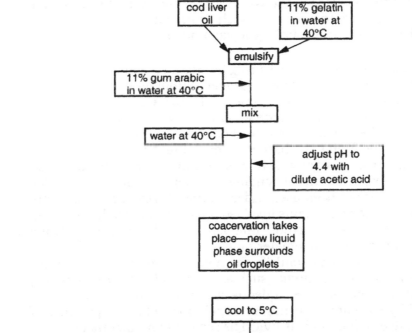

interaction of the charged polymers generates a separate phase that contains the ionic complex. The separate phase or coacervate must be of low viscosity and readily wet the dispersed nonpolar phase. Low initial viscosity has been found essential for spontaneous surrounding or engulfing of the nonpolar phase (Green and Schleicher 1957). There are now three coexisting phases: the continuous aqueous phase, the nonpolar phase that becomes the core, and a surrounding polymer-rich liquid phase. Continuous stirring is necessary to avoid agglomeration. The concentration of the polymer in the polymer-rich phase is then increased by manipulating the variables. For simple coacervation, the coacervate phase is increased gradually by further additions of alcohol or salt. This decreases the phase volume and increases the viscosity of the phase. For complex coacervation, the pH may be lowered further, which then requires solidification of the wall. In complex coacervation, the gelatin is usually cross-linked with glutaraldehyde. Cooling then further concentrates the polymer in the wall. If the particles are to be dried, it is often necessary to add solids such as fumed silica or talc to prevent the capsules from sticking to each other as they dry.

Interfacial Polymerization. During interfacial polymerization (Morgan and Kwolek 1959), the wall is created from monomers that may come from the two separate phases and polymerize at the interface (e.g., di- and triamines from the water phase and di- and triacyl chlorides from the organic phase [usually the dispersed phase]). They can also come from one phase if water is a reactant.

These processes are widely used in improving pesticides, making them less toxic and extending the time of effectiveness. The technique has not been used widely in the pharmaceutical industry (because the toxic reactants involved must be completely removed), but it is a valuable method uniquely suitable for encapsulating droplets in the low micrometer range.

The first step is to emulsify the nonpolar phase to be encapsulated (containing a monomer such as diacid chloride, di-isocyanate, or dianhydride) into the aqueous medium. This can be accomplished by mechanical stirring in a baffled vessel in the presence of a protective colloid such as PVA. The second monomer that is water soluble (e.g. a di- or polyamine) is then added (Ruus 1969). The formation of a polymer wall around the liquid droplet is very rapid and is usually complete within a few minutes. The reaction terminates because the wall itself isolates any excess monomer from the comonomer, which must come from the other phase. Generally some tri- or polyfunctional monomers must be present in both the aqueous and the nonpolar phases because of the high termination rate of the polymer chains under these conditions. Under this process, the walls are always cross-linked.

The two monomers do not always need to come from separate phases. Polyisocyanate from the organic phase can react with water, causing amines to form that then react at the phase boundary to produce polyurea walls. Larisch et al. (1994) discuss microencapsulation by interfacial polymerization of bacterial cells by minimizing the concentration of the most toxic reagents as well as choosing the right nonpolar solvents. Therefore, it is possible that interfacial polymerization may be applicable in the future to pharmaceutical processing.

Solvent Evaporation. This is a versatile process because it permits the formation of matrix particles (and occasionally microcapsules) from any polymer that can be dissolved in a solvent that has low solubility in aqueous solutions. A flowchart of the process is shown in Figure 4.5.

Figure 4.5. Solvent evaporation process.

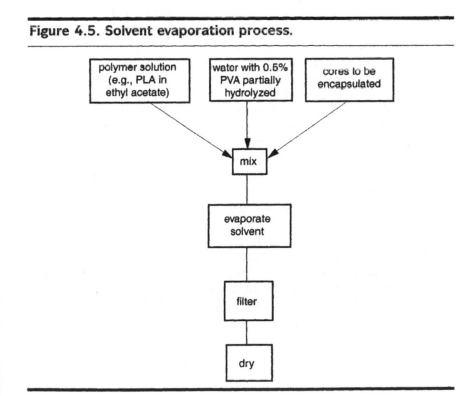

The active ingredient is either dissolved or dispersed in the polymer solution. The polymer solution is then dispersed in an aqueous dispersion medium containing a protective colloid such as PVA or hydrolyzed ethylene maleic anhydride (0.25 to 1 percent w/v). The droplets are solidified by removing the solvent by applying heat and blowing warm air through the suspension while mixing continues. The rate of evaporation is low and determines the time required. Even faster capsule formation might be achieved by spray drying or by stripping the solvent in a column. Solvent evaporation has been often used in the last few years in the preparation of controlled-release peptides, proteins, enzymes, and antigens for vaccines using poly-(D,L lactide) and lactide glycolide copolymers. In addition, many other polymers have been used: ethylene vinyl acetate (Elvax 40®) (Legrand et al. 1996), polystyrene, polymethyl methacrylate, ethylcellulose, polyvinyl chloride (Kentepozidou and Kiporissides 1995), polyurethane (Subhaga et al. 1995), and cellulose esters (Bhardwaj et al. 1995).

Gelation. If a sodium alginate solution (0.5 to 2 percent) is added dropwise to a calcium chloride or calcium acetate solution, droplets form immediately. Droplet formation is so rapid that the size of the droplets is generally conserved. If the droplets contain one or more oil droplets or a dispersed solid, even living cells, gelation produces a microcapsule or matrix particle. If the particles are retrieved and immersed in a polycationic polymer such as polylysine, an outer, interfacial, ionic polymer, complex membrane is formed. The core or bead can then be reliquified by treatment with a chelating agent, such as sodium citrate, which removes the calcium from the alginate but does not react with the polymer complex that forms the wall (Lim and Sun 1980). Outer walls can also be created by a photochemical process using acrylamide (Dupuy et al. 1988) or by reacting nascent carboxylic acid groups of polysaccharide esters with amino groups of proteins under alkaline conditions (transacylation) (Levy and Edwards-Levy 1996). Gels can also be formed with carrageenans or low methoxyl pectins by adding certain cations, particularly potassium or calcium salts, and chitosan by adding polyanions such as phosphate. Newer developments in microencapsulation by gelation include rapid photopolymerization of gels (Chandrashekhar et al. 1992; Sawhney et al. 1993).

Organic Continuous Phase

Polymer Phase Separation Via Polymer Incompatibility. Another type of phase separation or coacervation occurs when two solutions of different polymers, both soluble in the same solvent, are mixed. Two distinct phases are created, each phase being richer in one of the polymers (Figure 4.6) (Dobry and Boyer-Kawenoki 1947). This principle can be used to microencapsulate polar substances with water insoluble polymers (e.g., an aqueous phase or a drug). In practice, the polymer designed as the wall is a high molecular weight polymer with good physical properties (e.g., ethylcellulose). The polar liquid or solid to be encapsulated is dispersed in this high polymer solution. A lower molecular weight polymer (e.g., liquid polybutadiene) or even an oil, such as silicone or vegetable oil, is now added as the phase inducer (Bayless et al. 1972; Lapka et al. 1986). By the slow addition of the phase inducer solution, a second

Figure 4.6. Polymer-polymer phase separation process.

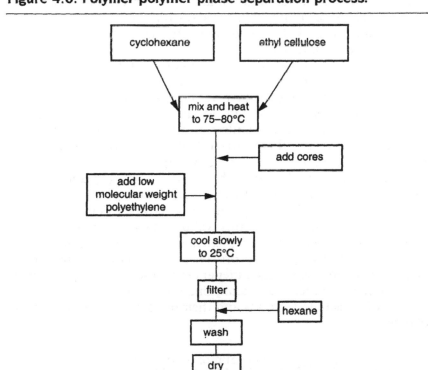

liquid phase rich in high molecular weight polymer is produced that preferentially wets the suspended polar droplet or drug core. This second phase must be of low viscosity so that it can spontaneously surround the droplet (Powell et al. 1968). By gradually adding more of the phase inducer, the volume of the second liquid phase or coacervate decreases and becomes stronger. Phase induction can also be strengthened by simultaneous cooling. Figure 4.6 summarizes a process that uses polymer incompatibility plus the effect of lowering the temperature. Before isolation, the microcapsules must be washed repeatedly with a nonsolvent for the wall polymer. During the drying of the microcapsules, finely divided solids such as starch, silica, or talc may be needed to prevent agglomeration.

SCALE–UP

The scale-up of a new chemical process is nearly always agonizing. After the process has been worked out in detail during development to give the correct products, the size of the equipment must be increased greatly to make large quantities of product at lower cost. Help in scaling up comes from three sources: dimensional analysis, an understanding of the underlying physics of the processes involved, and experimentation.

The subject of dimensional analysis on transport processes or unit operations is covered in chemical engineering textbooks and in many literature resources. It involves choosing the relevant properties, dimensions, forces, and rate variables for the processes being scaled, then forming dimensionless groupings of these variables. In a complete first listing, there can easily be 30 or more variables, giving rise to 25 or more dimensionless groups.

At first, this would seem to be of little help in guiding experimentation. However, many of the dimensionless groups will be ratios of properties such as temperatures, viscosities, densities, and so on. Most of the remaining groups will be ratios of rates characteristic of the basic processes, such as chemical reaction rates, flow rates, rates of heat transfer, and so on. These important groups can be used as dependent variables, independent variables, and parameters in planning experiments and analyzing the results.

The dimensionless groups and their physical significance are as follows:

$$\text{Reynolds Number} = \frac{\text{inertial momentum transfer}}{\text{viscous momentum transfer}} = \frac{\rho N D^2}{\mu}$$

$$\text{Power Number} = \frac{\text{power input}}{\text{inertial power}} = \frac{P}{\rho N^3 D^5}$$

$$\text{Froude Number} = \frac{\text{inertial force}}{\text{gravitational force}} = \frac{DN^2}{g}$$

$$\text{Weber Number} = \frac{\text{inertial force}}{\text{surface force}} = \frac{\rho N^2 D^3}{\sigma}$$

$$\text{Nusselt Number} = \frac{\text{convective heat transfer}}{\text{conduction}} = \frac{hD}{K}$$

$$\text{Fourier Number (Mass)} = \frac{\text{real time}}{\text{diffusion time}} = \frac{D_m t}{D^2}$$

where D is a characteristic length, such as vessel diameter or an impeller diameter; N is the rotation speed; P is the power input; g is the gravitational acceleration; h is the film heat transfer coefficient; k is the conductivity; t is the time; ρ is the density; μ is the viscosity, σ is the surface tension, and D_m is the diffusion coefficient.

In an ideal scale-up situation, the ratios of the important forces and rates would be the same in the small-scale apparatus and the commercial plant ("dynamic similarity"). However, this is possible only in simplistic cases. In the dimensionless groups above, the variables D and N occur in differing powers in several groups. Hence, if D and N are scaled to match the ratio of rates in one group, it is impossible to match the ratios represented by the other groups. Dimensional analysis alone does not solve scale-up problems; it is only a starting point. The most powerful tool is a detailed knowledge of the physics of the separate processes and an understanding of how they affect the prime goals of the scale-up. This helps to identify the more important dimensionless groups and the less important groups.

For example, if an exothermic reaction is occurring in an organic liquid emulsified in an aqueous carrier, the droplets may be small enough that conduction in the drops is very fast in both the small and the large process units. Hence, conduction in the drops will not cause difficulty in scale-up. However, as the system becomes larger, the ability to remove the heat generated by the reaction is likely to be very important, since the heat being generated increases as the cube of the dimension of the system, while the heat transfer area of the vessel wall increases only as the square of this dimension. A new design approach will likely be needed for the larger unit to maintain temperature control in the vessel.

Some dimensionless groups can be relegated to low priority by examining the assumption that the processing units on the two scales are geometrically similar. This is often not possible and, in some cases, is ridiculous. For example, if the goal of the process was to produce a microcapsule of a given diameter in both the small unit and the large unit, the particle diameter is fixed by this goal and is not to be scaled geometrically in the large unit. Hence, $D_{particle}/D$ is not a parameter in the scale-up. This is but one illustration of the fact that knowledge of the process and the goals is much more important than slavishly applying dimensional analysis.

Since most scale-up studies are not straightforward and the interaction of the various rates is neither clear nor simple, dimensional analysis is only a rough guide. Experimentation is usually required to be sure that the important goals are met as the system becomes larger.

Most microencapsulation processes have been scaled-up to production rates that are acceptable in the pharmaceutical industry. However, there are two important dimensional considerations that are important in these processes: (1) the practicality of the process as the microcapsule or particle diameter changes and (2) the problems that occur as the equipment size increases. These two considerations will be discussed below for the major microencapsulation processes.

Pan Coating

Tablets and large particles can be coated with ease in a pan coater, where gravity is the driving force for particle motion. The function of pan rotation is to move the bed of particles high enough to permit gravity to cause the particles to cascade downward, thus providing fresh particle surfaces for each contact with the spray and permitting proper drying of the coating without particle aggregation. As the particle diameter decreases, particularly below 2 mm, the particle-to-particle momentum that can be imparted by gravity decreases rapidly and soon becomes too slow to maintain a good downward motion of the particle bed. Aggregation begins to occur, forcing the spray rate to be decreased drastically, and the particle coating may become uneven. Nearly ideal conditions and an expert operator are required to place a good coating on particles smaller than 2 mm. In spite of the fact that pan coating seems simple and has been a useful process for many decades, balancing all of the variables to obtain an excellent product requires great skill. Careful automation is the best way to maintain a consistent product under these circumstances. A number of companies make excellent pan coaters that are widely used in the pharmaceutical industry.

Fluid Bed and Wurster Coating

Smaller particles can be coated easily in a fluid bed coater or a Wurster coater, because the requisite particle-to-particle momentum is provided by the energy in the fluidizing airstream. Much more air is used in these fluid bed processes than in pan coating to apply the same amount of a similar coating.

As the particle diameter is decreased, the fluidizing airflow (which is also the drying airflow) must be decreased to prevent the particles from blowing out the coating zone in the chamber. Since the drying rate becomes much slower, the spray rate of the coating must also be decreased. However, the surface area that must be coated increases as the particle diameter decreases. Hence, the time required for coating smaller particles increases rapidly as the particle diameter is decreased.

The major variable that must be controlled for smaller particle diameters is the mean size of the coating droplets being sprayed onto the particles. These droplets must be small compared to the core particles to avoid driving the particles together, causing aggregation or the particles to adhere to the inside of the cylinder.

The result of these increasing difficulties as particle diameter decreases is that these processes work well only on particles above roughly 100 μm in diameter; they can sometimes be used on smaller particles when the circumstances approach the ideal.

Wurster coaters and fluid bed coaters have been scaled successfully to hold roughly 500 kg of particles. Such equipment can coat several thousand tons of particles annually if the batch turnaround time is not long.

Spray Drying

Production rates for pharmaceutical products can be met with relatively small-capacity spray dryers. However, there are serious limitations on the particle size of the product that can be easily produced. A newer spray drying technique that requires a specifically designed fluid bed attached to the bottom of the drying chamber achieves granulation (agglomeration) of the small dry particles by forcing fine particles back into the drying zone.

When spray drying is being used to trap a flavor in a matrix particle, for example, the mean particle diameter of the dry particles will likely be in the range of 15 to 50 μm, and the particles will probably not be spherical. Atomizing the feed liquid into larger droplets is easy, but drying these large droplets becomes more difficult as the droplet diameter increases. This is due to the fact that the surface area–to–volume ratio is decreasing with increasing diameter, while the diffusion distance of water in the drop is increasing. In unsteady state drying, after the falling rate regime is reached (when water is no longer freely available at the surface), the drying rate becomes inverse to the square of the diffusion distance. Hence, the ability to dry a particle decreases more than quadratically with an increase in diameter. At the same time, as the droplet diameter increases toward 100 μm, the drop will tend to fall more vertically in the drying tower instead of following the swirling airstream. This decreases the residence time of the droplet in the drying air, further decreasing the amount of drying that can be achieved for that larger droplet. Hence, if droplet diameter is continually increased, a diameter will be reached where a small additional increase gives product particles that are not dry. A large dryer is required to properly dry droplets above 150 μm in mean diameter.

There are small bench top (laboratory-scale) spray dryers available from several manufacturers that have chamber diameters in the range of 10 to 20 cm. These dryers need only very small amounts of feed and are useful for initial testing, especially with expensive materials. However, they are limited to the formation of tiny product particles. Such dryers make it possible to observe whether a particular feed material will dry to a nontacky particle. To obtain information on the nature of the particle that might be made commercially, it is necessary to work with a dryer approaching 1 m in diameter in which a typical commercial atomizer can be used. For process information needed for scale-up, such as temperature as a function of spray rate, airflow rate, and solids concentration of the feed, data are usually required from even larger units.

Solvent Evaporation

The particle diameter of the matrix particles made by solvent evaporation covers an extremely wide range, from less than 1 μm to perhaps a 1 mm. This is because the first step in the process is the formation of an emulsion, with the droplets containing the active ingredient. The remainder of the process involves the drying and solidification of the emulsion droplets. Hence, virtually any emulsion that can be formed and prevented from aggregating during the drying process can be converted into tiny particles. However, in this slow process, if there is the least bit of tackiness to the droplets at any stage of drying, difficulty can be encountered in preventing aggregation of the smaller droplets.

This process cannot be scaled to high production rates (say, 1000 kg/day) in a straightforward manner. Laboratory work is carried out in stirred beakers, where it is not difficult to evaporate the solvent through the nonsolvent continuous phase in a reasonable period of time (e.g., 20–120 min). However, if a geometrically similar vessel were scaled from a 1 L beaker to a 100 L tank, the ratio of exposed surface to volume would drop by

$$\sqrt[3]{100}$$

meaning that it would take an exorbitant period of time to evaporate the solvent from the droplets. The design must be changed to provide more circulation to the surface of the tank, thus increasing the surface evaporation rate, or the design must move in the direction of a thin-film evaporator, which greatly increases the surface-to-volume ratio.

Coacervation Processes

Coacervation processes can easily handle the microencapsulation of droplets several millimeters in diameter. However, at the other end of the size spectrum, the smaller the droplet being coated, the more difficult it will be to prevent aggregation of the droplets.

One type of coacervation process has been scaled to batches as large as 10,000 L. In the scale-up, several distinct operations involving differing physical principles must be carried out on a large scale: emulsification to obtain the same droplet diameter, bulk mixing of continuously added liquid over the same time as in the small system, micromixing to achieve similar coacervate particle size with an even deposition of the coacervate onto the core droplets, and temperature control. The successful scale-up of this system by the precursors of Eurand America[*] can be counted as a significant achievement in scale-up.

[*]Eurand America, Vandalia, Ohio, was previously the National Cash Register Co. when the coacervation process was conceived and developed under the leadership of Dr. Barrett Green.

Spinning Disk Coating

The particle size limits in spinning disk coating are clearly defined. As the particle diameter is decreased, all of the variables must be adjusted to permit formation of a thin film coating on the disk between the particles compared to the diameter of the particle being coated. For coating viscosites that are low (e.g., 1–50 cp), it may be possible to coat particles as small as 20 μm. Even so, the fraction of coating on such small particles may be as high as 45 percent. As the core particle diameter increases, it is possible to form thinner coatings or to use more viscous coating liquids, even up to 5,000 cp for particles larger than roughly 700 μm.

The practical upper limit of the particle diameter that can be coated is not set by the process physics; it is set by cost. If the core particles are above 3–4 mm and the coating material is viscous (requiring higher rotational speeds), the particles will fly too far from the disk before the coating can be solidified before hitting the chamber wall. This means that the diameter of the coating chamber must be several meters. The cost of such large equipment may be prohibitive.

Part of the ease of scale-up of the spinning disk coating process is due to the fact that there is no scaling down of the coating disk for laboratory work. The disks used in the laboratory are virtually identical to the disks used in the large commercial plant. Hence, the physical parameters of the coating operation itself do not have to be scaled up. Although the process is capable of production rates similar to those of spray chilling, small units have been used successfully to coat pharmaceutical products.

PROBLEMS RELATED TO MICROENCAPSULATION

Problems Associated with Active Materials or Cores

Angular and Irregular Crystalline Shapes

Many pharmaceutical ingredients are crystals or crushed materials that have sharp edges and protrusions. They may even be needle shaped or shaped like plates. When such particles are coated, there will often be thinner coatings over the corners and protrusions, causing premature drug release. Therefore, thicker coats will be needed to achieve the desired protection for these irregular particles.

Rounded core particles can be coated more efficiently. Fortunately, there are several ways in which shape transformations can be accomplished. One can distinguish between granulation, crystal modification, pressure compaction, extrusion (Capes 1980; Shering-ton 1981; Perry 1984) and coating of the active constituent on round carrier particles. Particles granulated into larger particles will be rounder. Granulation can be carried out in fluid beds (Schoefer and Worts 1977), on rotating discs (Sparks et al. 1992), in spray dry-ers, and in wet-granulating mixers.

Melt granulation is the continuous melting and rapid solidifica-tion of the droplets of the active ingredient. If the substance has a relatively low melting point and is chemically stable at its melting point, melt granulation may be the best method. This method also has been successfully used in a partial melting process to granulate drugs that are unstable above their melting point. Granules can also be made by spraying a binder solution or melt into a fine powder of the substance in a fluidized bed or in a rotating pan, drum, or ves-sel equipped with mixing blades.

Crystal rounding can be done by abrading the sharp corners of crystals in crystallizers, as done with potassium chloride. Crystals can also be rounded by rapidly passing them through a hot zone that is designed only to melt the outer layer. Rounded shapes can be produced by pressure compaction similar to a tableting operation and by agglomeration in a rotating drum. Spraying the active ingre-dient onto nonpareils is widely practiced in fluidized beds and pan coaters. This works well if the dose is low or the agent is very po-tent.

Physicochemical Changes

Changes in crystallinity or morphology during processing can be a serious problem. The typical drug or organic chemical is polymor-phic (i.e., it can exist in several crystalline forms or morphologies). This is particularly common in compounds that have long hydro-carbon chains relatively free of strong polar groups. For instance, compounds as simple as triglycerides have three or four crystalline forms (Ekey 1954). Usually, one of the forms is the most stable. It is possible that the process transforms the active substance or an ex-cipient to a higher energy, less stable form. This form can later slowly change to a more stable form. If properties (e.g., solubility) differ for each of the crystalline forms, release rates may differ, and bioavailability could change with time. How this is best dealt with depends on the system. The formulation can be changed or the product annealed to speed up the transformation.

High Osmotic Pressure of Low Molecular Weight, Highly Soluble Substances

When oral dosage forms are immersed, a water gradient is immediately established across the coating. Water will be transported across the wall at a rate consistent with its diffusion coefficient and solubility in the wall. The water will then begin to dissolve the active substance. The pressure inside the microcapsule will then increase. What happens next will depend on the properties of the wall. If the wall is very strong, the pressure will rise and stress the wall to make it expand, increasing its permeability. A pseudosteady state can be reached where the rate of water drawn in by the chemical potential just balances the rate at which it is pumped out by the higher pressure inside. The Alza Corporation (Palo Alto, CA) osmotic pump makes use of this phenomenon by providing a small hole where the saturated solution can leave at a nearly constant rate. If the wall is brittle and the pressure of the solution inside the microcapsule exceeds its strength, the wall will rupture and the drug will be released rapidly (Cheng 1994).

Wide Size Distribution of Core Materials

Typical solid USP materials come in particle size distributions such that d_{60}/d_{10} ranges from 1.5 to as high as 5; d_{60} is the diameter corresponding to the size where 60 percent of the weight of the material is smaller; d_{10} is the diameter where 10 percent of the material is smaller than the stated size. Such broad size distributions sometimes cause small particles to be incorporated into the wall of larger particles. The mixing regime or the airflow that may be ideal for the largest particles may be entirely too rapid for smaller particles. This always restricts the type of release curve that can be obtained. In theory, only a monodisperse population of microcapsules can give a zero-order release (Dappert and Thies 1978).

Problems Caused by Limitations of Wall Materials

Insufficient Choice of Materials Approved for Pharmaceuticals and Foods

The limited materials available is immediately apparent by comparing the number of water-insoluble polymers approved for pharmaceutical coatings with all commercial polymers. Many polymers would be entirely inert in the gastrointestinal tract, and one would think it would be sufficient to prove that solvents and monomers are present at sufficiently low concentrations. However, entire classes of processes (e.g., interfacial polymerization) are virtually ruled out for both foods and pharmaceuticals. Many substances that are approved (e.g., fats and waxes) are too weak and brittle to serve as good walls for highly soluble substances.

Adherence and Stickiness Problems of Wall Materials During Processing

Most microencapsulation operations in which either water or solvents are used suffer from recovery problems because of the stickiness of the polymer walls during recovery and drying. Therefore, it is often necessary to use a fine powder, such as fumed silica, talc, or starch, to reduce stickiness. During fluid bed coating, there is often a buildup of static charges, which causes particles to stick to each other temporarily if there is insufficient humidity. The charge buildup comes from the motion of particles against internal surfaces. If the particles do not circulate properly, the coating is distributed unevenly.

Permeability of Wall Materials

The permeability of acceptable wall materials is such that, with any organic wall, long-term protection is not possible against water or oxygen for particles below 100 μm. Long term means more than 1 h, if the wall is on the same order of thickness or less than the particle diameter. For controlled release from particles of substances of high water solubility, only polymer coatings or walls such as ethylcellulose are satisfactory. For short-term taste masking, fats and waxes reinforced by polymers may be adequate. Plasticizers that may be required to increase the flexibility of the film for tableting also increases the permeability of polymers.

Problems Caused by the Limitations of the Processes

Nonlinear Nature of Most Processes

Wall thickness and permeability are generally not linear functions of the process variables. This is particularly true for microencapsulation processes that occur in continuous liquid phases. The range of the variables that produce useful products may also be narrow.

Interactions Between Walls, Cores, and Processing Materials

In microencapsulation processes with reactive monomers, such as interfacial polymerization, the monomers may also react with the active substance. With coacervation processes, the wetting of the core is very important. Small amounts of surfactants or trace impurities can change the wetting characteristics of the cores. Polymer walls deposited from different solvents will have different morphologies and different permeabilities.

Difficulties in Obtaining Zero-Order or Constant Release

Careful examination of microcapsules shows that there no two microcapsules are alike. This can be illustrated by encapsulating a pH indicator. The capsules are then immersed in a drop of a solution of a different pH and observed under the microscope. If the color change of the individual capsules is timed, it is found that the time of color change can differ by several orders of magnitude from one microcapsule to another, even if they initially appear identical.

Microcapsules from a typical process often give a release curve that may be called pseudo first order. The rate of release is proportional to the amount remaining inside. The logarithm of the amount still retained versus time gives a straight line, which represents the sum of the release of all particles. Some particles empty rapidly; others empty slowly. For zero-order release from microcapsules to be possible, a monodisperse population with respect to size and permeability is necessary. Both the drug load and the size need to be very uniform, which is usually quite difficult to obtain.

Process Parameters Influencing Product Quality

The process parameters that must be controlled depend on the particular process under consideration. For fluid bed coating, the following need to be controlled and monitored in order to obtain a reproducible product:

- The rate of pumping of the coating.

- The air pressure to the nozzle.

- The liquid pressure on the nozzle air (which can be used to indicate plugging of the nozzle).

- The temperature, flow rate, and humidity of the fluidizing air.

- The temperature of the bed and the exit air.

- The pressure drop across the bed and across the filters.

For coacervation, the following need to be controlled and monitored:

- pH.
- Temperature of the batch.
- Rate of stirring.
- Rate of addition of the ingredients.
- Size of the droplets and particles.
- The temperature of the heating/cooling jacket.

For the spinning disk coating process, the following should be monitored:

- The rate of disk rotation.
- The temperature of the coating stream.
- The feed rate of the particles to be coated.
- The feed rate of the coating stream.
- The temperature of the disk or the power to the disk.
- The temperature of the cooling air supplied and the rate of supply of the cooling air.

Mass and Heat Balances

Mass and heat balance calculations should be made on all processes, if possible, since this procedure often leads to better processing and improved economies.

The Consistency of the Release Behavior

Batch-to-batch consistency or the initial release profile as compared to the profile after extended storage must be considered. Careful identification and monitoring of critical parameters affecting barrier performance are the most important. This is most critical when naturally occurring raw materials, with their inherent variability, are used, since batch-to-batch variation can affect product performance. Shellacs, and to some degree CAP coatings, can exhibit a slowing of release characteristics due to cross-linking reactions during storage. Over storage lifetimes, some materials, such as triglycerides, may transform among crystalline forms that have differing permeabilities to biological fluids.

Batch-to-batch consistency problems can be related to process parameters (e.g., spray rates, humidity control, process temperatures, stirring rates, and feed rates), problems in raw material supply, or parameters that are not thoroughly understood. The remedy can sometimes be as simple as posttreatment drying or heating or an extra washing step. For all processes, a reduction in the art needed and an increase in the scientific understanding of the processes generally will improve the consistency of results.

REFERENCES

Akbuga, J., and Bergisadi, N. 1995. 5 flurouracil-loaded chitosan microspheres: Preparation and release characteristics. *J. Microencaps.* 13 (2):161.

Arnaud, P., C. Boué, and J. C. Chaumeil. 1995. Cellulose acetate butyrate microparticles for controlled release of carbamazepine. *J. Microencaps.* 13 (4):407.

Arneodo, C., J. P. Benoit, and C. Thies. 1986. Characterization of complex coacervates used to form microcapsules. *S.T.P. Pharma* 2 (15):303.

Baker, J. P., and R.A. Siegel. 1995. Poly (n-isopropylacrylamide) hydrogel membranes for pulsatile drug delivery. In *Proceed. of the Inter. Symp. Control. Rel. Bioact. Mater.* 22:340.

Banker, G. S. 1966. Film coating theory and practice. *J. Pharm. Sci.* 55:81.

Bayless, R. G., C. P. Shank, R. A. Botham, and D. Werkmeister. 1972. Process of forming minute capsules and three-phase capsule-forming system useful in said process. U.S. Patent 3,674,704 to National Cash Register.

Bhardwaj, W. B., A. J. Shukla, and C. C. Collins. 1995. Effect of varying drug loading on particle size distribution and drug release kinetics of verapamil hydrochloride microspheres prepared with cellulose esters. *J. Microencaps.* 12 (1):71–81.

Brannon-Peppas, L., and N. A. Peppas. 1990. Dynamic and equilibrium swelling behaviour of pH-sensitive hydrogels containing 2-hydroxyethyl methacrylate. *Biomaterials* 11:635.

Brannon-Peppas, L., and N. A. Peppas. 1991. Time-dependent response of ionic polymer networks to pH and ionic strength changes. *Intern. J. Pharmac.* 70:53.

Bungenberg de Jong, H. C. 1949. Complex colloid systems. In *Colloid Science,* vol II, edited by H. R. Kruyt. New York: Elsevier.

Capes, C. E. 1980. *Particle size enlargement.* New York: Elsevier.

Chandrashekhar, P., J. A. Hubbell, et al. 1992. Rapid photopolymerization of immunoprotective gels in contact with cells and tissue. *J. Amer. Chem. Soc.* 114:8311–8312.

Chatfiled, H. W. 1962. *Science of surface coatings.* New York: Van Nostrand, p. 453.

Cheng, P. 1994. A systematic study of osmotic rupture and the corresponding drug release. Ph.D. Thesis, Washington University, St. Louis, Mo., USA.

Cleland, J. L. and A. J. S. Jones. 1995. Development of stable protein formulations for microencapsulation in biodegradable polymers. *Proceed. Intern. Symp. Control. Rel. Bioact. Mater.* 22:514.

Daniels, R., and E. M. Mittermaier. 1995. Influence of pH adjustment on microcapsules obtained from complex coacervation of gelatin and acacia. *J. Microencaps.* 12 (6):591.

Dappert, T., and C. Thies. 1978. The heterogeneous nature of microcapsules. *Proc. 5th Int'l. Sym. on Controlled Release of Bioactive Materials.* Deerfield, Ill., USA: Controlled Release Society, Inc.

Deasy, P. B. (1984) *Microencapsulation and related drug processes.* New York: Marcel Dekker, pp. 39, 161–180.

Dobry, A., and F. Boyer-Kawenoki. 1947. Phase separation in solution. *J. Poly. Sci.* 2:90–100.

Donbrow, M., and M. Friedman. 1974. Permeability of films of ethyl cellulose and PEG to caffeine. *J. Pharm. Pharmacol.* 26: 148–150.

Dupuy, B., H. Gin, C. Baquey, and D. Ducassou. 1988. In situ polymerization of a microencapsulating medium round living cells. *J. Biomed. Mater. Res.* 22:1061–1070.

Ekey, E. W., ed. 1954. *Vegetable fats and oils.* ACS Nomograph #123. Washington, D.C.: American Chemical Society, pp. 108–132.

Gao, P., J. W. Skoug, P. R. Nixon, T. R. Ju, N. L. Stemm, and K. C. Sung. 1996. Swelling of hydroxypropyl methyl cellulose matrix tablets 2. Mechanistic study of the influence of formulation variables on matrix performance and drug release. *J. Pharm. Sci.* 85:732–740.

Green, B. K., and L. Schleicher. 1957. Oil containing microscopic capsules and method of making them. U.S. Patent 2,800,457.

Hamid, I. S., and C. H. Becker. 1970. Release study of sulfaethylthiadiazole from a tablet dosage form prepared from spray congealed formulations of SETD and wax. *J. Pharm. Sci.* 59 (4):511.

Jeffery, H., S. S. Davis, and D. T. O'Hagan. 1993. The preparation and characterization of poly(lactide-co-glycolide) microparticles. II. The entrapment of a model protein using a (water-in-oil)-in-water emulsion solvent evaporation technique. *Pharm. Res.* 10 (3):362–368.

Kentepozidou A., and C. Kiporissides. 1995. Production of water-containing polymer microcapsules by the complex emulsion/solvent evaporation technique: Effect of process variables on microcapsule size distribution. *J. Microencaps.* 12 (6):627–638.

Khalil, S. A. H., J. R. Nixon, and J. E. Carless. 1968. Role of pH in the coacervation of the systems gelatin-water-ethanol and gelatin-water-sodium sulfate. *J. Pharm. Pharmacol.* 20:215.

Lapka, G. G., N. S. Mason, and C. Thies. 1986. Process for preparation of microcapsules. U.S. Patent 4,622,244.

Larisch, B. C., D. Poncelet, C. P. Champagne, and R. J. Neufeld. 1994. Microencapsulation of *Lactococcus lactis* subsp. *cremoris*. *J. Microencaps.* 11 (2):189–195.

Lee, H. K., J. H. Park, and K. C. Kwon. 1996. Double walled microparticles for HBV single shot vaccine. *Proceed. Intern. Symp. Control. Rel. Bioact. Mater.* 23:333.

Lefebvre, A. H. 1989. *Atomization and sprays*. Washington, D.C.: Hemisphere Publishing.

Legrand, J., L. Brujes, G. Garnelle, and P. Phalip. 1996. Study of a microencapsulation process of a virucide agent by a solvent evaporation technique. *J. Microencaps.* 12 (6):639–649.

Levy, M. C., and F. Edwards-Levy. 1996. Coating alginate beads with cross-linked biopolymers: A novel method based on a novel transacylation reaction. *J. Microencaps.* 13 (2):169–183.

Lim, F., and A. M. Sun. 1980. Microencapsulated islets as bioartificial endocrine pancreas. *Science* 210:908–910.

Maa, Y. F., C. Hsu. 1996. Liquid-liquid emulsification by static mixers for use in microencapsulation. *J. Microencaps.* 13 (4):419–433.

Marshall, W. R. 1954. *Atomization and spray drying.* American Institute of Chemical Engineers, Monograph Series No. 2. 50:12–49.

Masters, K. 1979. *Spray drying handbook,* 3rd ed. New York: Halsted Press, pp. 165–290.

Masters, K. 1991. *Spray drying handbook,* 5th ed. New York: Halsted Press, pp. 643–662.

Merrington, A. C., and E. G. Richardson. 1947. The breakup of liquid jets. *Proc. Phys Soc.* 59 (1):1–13.

Morello-Mathias, P. A. 1995. Zero order release from a brittle matrix due to osmotically-induced surface erosion. Ph. D. thesis, Washington University, St. Louis, Mo., USA.

Morgan, P. W., and S. L. Kwolek. 1959. Interfacial polycondensation. II. Fundamentals of polymer formation at liquid interface. *J. Poly. Sci.* 40:299–327.

Morse, L. D., and P. A. Hammes. 1975. Ethyl cellulose encapsulated nutrients. U.S. Patent 3, 860, 733.

Müller, B. W., and F. Yunis-Specht. 1994. *Proceed. Intern. Symp. Control. Rel. Bioact. Mater.* 21:762.

Perry, R. H. et al. 1984. *Perry's chemical engineer's handbook,* 6th ed.

Porter, S. C., R. J. Haluska, et al. 1991. A Seminar on Film Coating Technology presented at Colorcon: "Processing Aspects of Conventional Film Coatings," pp. IV-1 to IV-33.

Powell, T. C., M. E. Steinle, and R. A. Yoncoskie. 1968. Microencapsulation process. U.S. Patent 3,415,758.

Rayleigh, L. 1879. On the instability of jets. *Proc. Lond. Math. Soc.* 10 (4):4–13.

Remunan-Lopez, C. M. J. Alonso, T. Calleja, J. L. Villa-Jato, and R. Bodmeier. 1995. Development of new chitosan-gelatin microcapsules by complex coacervation process. *Proceed. Intern. Symp. Control. Rel. Bioacti. Mater.* 22:430.

Rowe, R. C. 1986. The effect of the molecular weight of ethyl cellulose and hydroxypropyl methyl cellulose. *Int. J. Pharm.* 29:37.

Ruus, H. 1969. Method of encapsulation. U.S. Patent 3,429,827.

Sakr, A., and J. R. Oyola. 1986. Some factors affecting the dissolution of microencapsulated potassium chloride in directly compressed tablets. *Pharm. Ind.* 48:1.

Samuels, W. E., and R. E. Sparks. 1973. Fluidropper: A device for forming highly uniform drops. *Rev. Sci Instr.* 44 (2):132–134.

Sawhney, A. S., J. A. Hubbell, et al. 1993. Bioerodible hydrogels based on photopolymerized poly(ethylene glycol)-copoly(alpha-hydroxy acid) diacrylate macromers. *Macromolecules* 26:581–587.

Schoefer, T., and O. Worts. 1977. Control of fluidized bed granulation. *Arch. Pharm. Chemi Sci. Ed.* 5:178–193.

Scott, M. W. et al. 1964. Spray congealing: Particle size relationships using a centrifugal wheel atomizer. *J. Pharm. Sci.* 53 (6):670.

Sherington, O. 1981. *Granulation.* London, UK: Heyden.

Shinar, R. 1961. On the behavior of liquid dispersions in mixing vessels. *J. Fluid Mech.* 10: 259–275.

Shinar, R., and J. M. Church. 1960. Predicting particle size in agitated dispersions. *Ind. Eng. Chem.* 52 (3):253–256.

Siekmann, B., and K. Westesen. 1994. *Colloids Surfaces B: Biointerfaces* 3:159.

Sparks, R. E., and N. S. Mason. 1987. Method for coating particles or liquid droplets. U.S. Patent 4,675,140.

Sparks, R. E., N. Mason, and M. Center. 1992. Method and apparatus for granulation and granulated product. U.S. Patent 5,100,592.

Subhaga, C. S., K. G. Ravi, M. C. Sunny, and A. Jayakrishnan. 1995. Evaluation of an aliphatic polyurethane as a microsphere matrix for sustained theophylline delivery. *J. Microencaps.* 12 (6):617–625.

Summerville G. R., Jr. 1962. Apparatus for hardening by cooling or chemical reaction. U.S. Patent 3,015,128.

Uchida, T., K. Yoshida, and S. Goto. 1996. Preparation and characterization of polylactic acid microspheres containing water-soluble dyes using a novel w/o/w emulsion solvent evaporation method. *J. Microencaps.* 13 (2):219.

Yoshida, R., T. Okano, Y. Sakurai, and K. Sakai. 1994. Control of "on-off" switching temperature in pulsatile drug release using n-isopropylacrylamide copolymer gels. *Proceed. Intern. Symp. Control. Rel. Bioact. Mater.* 21:646.

5

AUTOMATED
COATING SYSTEMS

Gregory Raymond Smith
James G. Spencer

Vector Corporation

This chapter is divided into five sections. The first section provides a brief historical account of the different methodologies used for automated coating processes. Each control methodology will be discussed as follows:

- Evolution of the methodology.

- Basic principles behind the methodology.

- Applicability to the design of modern automated coating systems.

The second section of the chapter presents the elementary aspects of any tablet-coating system and how these aspects are automated. The section will illustrate how automated control systems are defined, present the terms *hardware* and *software* in relation to automated coating systems, describe the advantages and disadvantages of automated coating systems, and define how the quality of automated coating systems can be evaluated.

The third section focuses primarily on designing a high-quality software system for automated coating systems. Since software is such an important part of any automated coating system, anyone responsible for specifying, operating, designing, and maintaining these systems needs to understand the basic principles of software development. The software life cycle will be presented in order to define how quality is built into software systems.

The fourth section focuses on the operation of a well-designed automated coating system. It will define the basic requirements of operating a coating system, the different automation processes that occur with an automated coating system, and define the role of supervisory stations and their effectiveness in monitoring or controlling automated coating systems.

The final section discusses the concept of batch control standardization through standard ANSI/ISA-S88.01. It focuses on the specific impact of this standard on the pharmaceutical industry.

HISTORICAL METHODS OF AUTOMATED COATING SYSTEMS

In the past 20 years, the broad field of industrial controls has experienced a tremendous amount of technological advancement. These advancements are analogous to the tremendous growth in manufacturing technologies during the Industrial Revolution. Many of these technological improvements have filtered their way into the specific field of automated coating systems. Not all industrial control advances were applicable to tablet-coating systems; however, many of these advances revolutionized the design of automated coating systems.

This section will briefly present the history and theory behind the most influential control methods and will describe how each method contributed to automating coating systems.

Pneumatic Control

Some of the first automated coating systems consisted of devices that were controlled pneumatically. Controlling a device "pneumatically" simply meant that the amount of air pressure supplied to a device was varied in order to achieve a desired response or output. For example, a pneumatic sensor could be used to measure the temperature of process air entering a tablet coater. This sensor would vary the amount of air pressure exiting the device in relation to a change in process air temperature. Many of the first automated coating systems consisted of pneumatically controlled devices.

Pneumatic control is still a popular method. Since the delivery of a coating solution to the tablet bed typically requires clean, compressed airflow through spray guns, instrument quality air (air that is clean enough to operate pneumatic devices) is already available for most coating applications. Further, the air pressure required to deliver the coating solution is the same pressure required to operate most pneumatic devices (50–90 psi).

There are coating applications where pneumatically controlled devices are the safest control devices available. Coating solutions often use acetone or other hazardous materials that can, under certain circumstances, be explosive. When dealing with explosive materials, pneumatically controlled devices are often preferred since there is no potential energy stored in the device that could create a spark that would precipitate an explosion.

However, there are several inherent problems with pneumatic control, which make the technology difficult in designing complex automated systems. The types of pneumatic control devices used for automated coating systems are limited to simple time-delay devices (important for automatic sequences) and basic alarm-monitoring systems.

Another drawback to pneumatic control is the limited precision as compared to electrical methods of automating coating processes (discussed later in this section). In a pneumatic temperature sensor, for example, the air pressure from the device changes according to changes in the air temperature. The variance in air pressure is displayed on a pneumatic gauge by a needle that rotates freely on a dial in relation to air pressure. The dial is graduated so that the operator can determine the process air temperature (or other process variable). These gauges are still the primary source of inaccuracy for pneumatically controlled devices. Operators looking at the dial from different angles can often record different readings. Many times, the accuracy of a pneumatic gauge is dependent on the thickness of the needle, which may span as much as ±10 percent of the full-scale reading of the dial.

Inaccuracy also results from the amount of pressure loss due to the resistance of the pneumatic tube in which the air pressure is transferred. When designing pneumatically controlled systems, the length of pneumatic tubing that transfers air signals from one point to another must be minimized. The longer the pneumatic tube, the more the pneumatic signal will be reduced due to losses in that tube.

Instrument quality air is always required to operate pneumatic devices. Pneumatically controlled devices are always subject to failures if the air supplied to these devices is contaminated or dirty. The speed of pneumatically controlled devices is also associated with the length of the pneumatic tube. Even with very short pneumatic tube lengths, the speed of pneumatic signal transmission is very slow. This speed difference is even more apparent when compared to another method of automatic control on early coating systems: electromechanical control.

Electromechanical Control

An electromechanical device harnesses the potential energy of electricity and converts it into mechanical energy. The electric motor, critical in any coating application, is a prime example of the amount of mechanical energy that can be harnessed through electromechanical devices.

There are other inherent advantages in using electromechanical devices. Since electricity travels at the speed of light (Nillson 1987), electrical devices are certainly faster than their pneumatic counterparts. Electricity is transmitted from point to point by using conductive metal wires (copper, aluminum, etc.) rather than air tubes as required for pneumatic control. Metal wiring is much smaller and easier to install than pneumatic tubing. Also, the length of the metal wiring could be much greater than the length of pneumatic tubing because electrical losses from wiring are much less than pressure losses from tubing.

Yet, there are some drawbacks to electromechanical devices. Electricity is capable of producing sparks that could generate an explosion if a flammable solution is used for the tablet-coating process. The entire field of "intrinsic safety" is based on the premise of limiting the amount of electrical energy in a circuit, such that an explosion-producing spark cannot be generated. Intrinsically safe devices are often more expensive than their pneumatic counterparts.

All electrical devices emit electromagnetic fields. These fields can actually alter or disrupt other electrical devices that may be in close proximity. The European Community pays particular attention to this problem. In fact, it has established electromagnetic standards (the EMC Directive [EC 1995]) to limit the amount of radiation that an electrical device (or system) can radiate and the amount of immunity that the device (or system) has from other electromagnetic fields.

Besides the electric motor, the most influential electromechanical device used for automated coating systems is the control relay—the backbone of automated coating systems in the 1970s and early 1980s. The concept of a control relay is based on the fact that a complete electrical circuit must be established in order for an electrical device to become activated. The control relay will either connect or disconnect an electrical circuit in order to control the "on/off" state of another device. There are only two states in which a control relay can operate: "open" or "closed." If a control relay is "open," the electrical circuit is disconnected; any device attached to that circuit is then deactivated. The reverse is true if a control relay is "closed." For example, consider the simple schematic shown in Figure 5.1.

By opening or closing the control relay, the motor can be deactivated or activated as part of an automatic sequence. Designers of automated coating systems control the state of control relays in order to achieve a certain amount of automation. For example, designers build control systems such that a control relay (or several control relays sequenced together) would be activated according to an automatic coating sequence. Many coating systems utilizing electromechanical control would have several control relays associated with the system. In fact, the term *ladder diagram* is often used to describe complicated schematics, since several relays shown on a piece of paper look like ladders stacked side by side. Appendix A shows a schematic from a section of an automated coating system that utilizes electromechanical devices.

The operator of a coating system with electromechanical devices typically would use standard push buttons mounted on a control panel to activate and deactivate devices associated with the coating process. These control panels are often very large and may have 20–40 (or more) push buttons that must be pressed at some time during the process. Even though it is a cumbersome technology, electromechanical control systems have a market niche today—

Figure 5.1. Control relay used for automatic control of a motor.

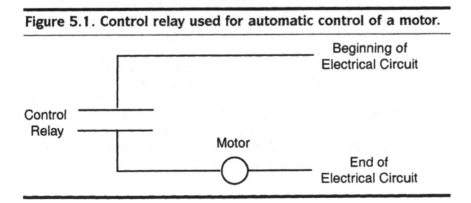

primarily with customers who have very simple automated coating processes.

The advantage of electromechanical control is that automated coating systems can be "programmed" according to the amount of control relays that are part of the system. However, there are some disadvantages to this control methodology. Control relays require a large amount of panel space. In order to design systems with a limited amount of functionality, many control relays are required. The control panels that house the control relays can span several feet in length.

Another disadvantage is that in order to change the "programming" of the control system, the control panel must be rewired to modify the actions of the control relays. Oftentimes, rewiring the control panel to change a simple part of the coating sequence takes a tremendous amount of time. Thus, electromechanical control limits the amount of automation that can be part of the coating process. Another technology, the programmable logic controller (PLC), emerged to solve many of the problems associated with electromechanical control devices.

Digital Control and the Programmable Logic Controller

In the 1970s and 1980s, electrical devices in the automation of coating systems increased steadily. The concept of digital control (the technology used in personal computers [PCs] today) gave birth to devices such as digital displays. These displays replaced analog gauges that were prevalent with pneumatically controlled devices. Independent process controllers, another by-product of digital control, were introduced to modify automatically the actions of one device in order to achieve a desired response in a process variable (e.g., modifying the position of a damper inside the ductwork of the tablet coater in order to achieve a desired airflow). However, the most revolutionary device that originated from the digital control methodology was the PLC (Wilbanks 1996).

The PLC was first introduced in the automotive industry in the late 1960s and early 1970s. PLCs were introduced to the coating industry much later because of its high initial costs. The first PLC functioned very similar to control relays; however, several electrical components (from multiple electrical circuits) could be controlled from a single PLC. The PLC would activate and deactivate these components according to a software program that was stored inside its memory.

As the PLC became more cost-effective, it gradually replaced many electromechanical control systems (Wilbanks 1996). The main benefits of the PLC are as follows:

- Modifications to the PLC's software program are made in a time-effective manner. Changes occur at the touch of a few keystrokes from a programming terminal—rewiring is not required.

- The space requirements for a PLC are much smaller than for an equivalent system composed of control relays.

- All of the system's components (temperature sensors, airflow sensors, motors, valves, etc.) can be connected to the PLC for complete control of the coating process.

- The cost of purchasing and programming a PLC is less than the cost of purchasing a large amount of relays and wiring these devices into a control panel.

- Complex mathematical expressions can be calculated inside the PLC, thus allowing the PLC to have internal proportional, integral, and derivative (PID) control for key aspects of the coating process (such as air temperature and airflow) (Considine 1985). PID control is a method that utilizes complex mathematical expressions to vary the output of a control device (such as a damper mounted in the ductwork) in order to control a process variable to a desired level (the amount of airflow in a tablet coater). PID control loops are discussed in the next section of this chapter.

- Most PLCs can store data, which is very important to automated coating systems. The process setpoints now can be stored inside the PLC. As the amount of storage capacity inside the PLC becomes larger, several recipes of different coating processes can be stored within the PLC.

The PLC also spurred tremendous growth in the area of operator interface terminals (OITs). The PLC did not have an effective method of viewing the information that was stored inside the PLC (such as the current value of process variables, process setpoints, and alarm information). An OIT is typically a digital display device (i.e., monitor) that reflects the current status of the devices controlled by the PLC and allows an operator to monitor and/or control the coating process. The first OITs simply had small text displays that allowed for a single process variable to be displayed from the PLC. As the PLC evolved, so did the OITs. Today, OITs have complete graphical capabilities, come in various sizes, and can communicate with a variety of different PLCs. The more sophisticated OIT has push buttons integrated into the terminal or utilizes a touch screen so an operator can control the coating process. Since the OIT is the main interface for the operator of an automated coating system, the sophistication level of the OIT often determines the amount of automation that is associated with a coating process.

Because of the advantages of PLCs, manufacturing facilities require PLCs throughout the plant to perform various functions and interface with other areas of the facility. This is known as distributed process control (as compared to centralized process control where one processor controls the entire plant). Distributed process control is advantageous because if one processor fails, the entire plant is not shut down.

Digital Control—Distributed Control Systems

Early PLCs had many limitations, including limited amounts of programming commands, devices that could be connected to the PLC, and memory to store the software program. These limitations made distributed control with PLCs unmanageable and cost prohibitive for most manufacturers wanting to implement large-scale control systems. In the mid-1970s, the first distributed control system (DCS) was introduced to the pharmaceutical industry by Yokogawa in Japan and Honeywell in the United States (Wilbanks 1996).

The DCS was designed to provide a more cost-effective and feasible alternative to large-scale control systems. Its development highlighted the benefits of distributed process control. The DCS had the same advantages as the PLC, except that the costs to install a DCS were prohibitive for smaller coating applications. Consequently, the application of DCSs to automated coating systems was limited. Manufacturers that tended to use DCS technologies for tablet-coating processes were companies that had plant-wide automation system installed in their facilities.

Distributed control offers manufacturers the ability to view processes from a remote location. Since information about the process must be fed into the DCS, it would be relatively simple to retrieve that information from the process room and display it at a remote location. Supervisory control and data acquisition (SCADA) systems would be considered an offshoot of the DCS.

The boundaries between DCSs and PLCs are more blurred today. The price of DCSs are significantly lower today than when they were first introduced. Furthermore, PLC manufacturers have increased the amount of functionality and storage capacity associated with their devices. There have been many debates as to which control philosophy is better for tablet-coating applications; however, these arguments are beyond the scope of this text.

Digital Control—Personal Computers

The PC offered coating manufacturers an alternative to the PLC and the DCS. However, the PCs in the late 1980s and early 1990s were very slow in process-controlling applications when compared to PLCs and DCSs. PCs were slower than PLCs and DCSs because PLCs and DCSs had processors that were designed specifically for controlling and monitoring electrical devices. PLCs and DCSs were simply too fast for the PC. In the last few years, PCs have become orders of magnitude faster than what they were even five years ago. The speed difference between the PC and the PLC/DCS is diminishing, thus making the PC more attractive. There is an increasing amount of programs ("soft PLC" programs) on the market today that make a PC operate in the same manner as a PLC. Improvements to PCs (primarily in networking and storage capabilities) opened the door for another control methodology for modern coating systems: supervisory control and data acquisition systems.

Supervisory Control and Data Acquisition Systems

In the late 1980s to early 1990s, SCADA systems became available for automated coating processes (Wilbanks 1996). These systems operated on PCs and typically did not perform the same functions as the PLC. SCADA systems still required PLCs to control the electrical devices that comprised the coating system. PLCs were required because they were typically faster than the PC in controlling electrical devices. However, the SCADA system could act as the operator interface for the coating system and allow the operator to monitor and control all aspects of the coating process.

SCADA systems were potentially much more than simple OITs. In fact, SCADA systems, in conjunction with PCs, revolutionized the automation applied to today's coating systems. Since the SCADA system ran on PCs, the networking capabilities developed for the PC market were applied to coating processes. Prior to SCADA systems, most networking between PLCs and OITs was based on proprietary networking standards developed by PLC and/or OIT manufacturers. Supervisory or remote monitoring stations could now be connected via ethernets (or any other standard networking method) instead of proprietary networks. This paved the way for a more standard method of connecting several SCADA systems together.

The improvements made to the storage capabilities of PCs allowed SCADA systems to store a vast amount of information that could be accessed by any other SCADA system on the same network. The amount of information stored on the PC was much greater than equivalent PLC and DCS systems at the same cost. Historical process information could be stored on the PC and retrieved by another SCADA system. Recipes (a collection of setpoints and process steps that govern the coating process) could be created, maintained, and used for future coating applications.

Future Control Methodologies in Coating Processes

The future of automated coating processes lies in the advancements of the PC and the Internet. Improvements to the PC will continue and eventually will replace the PLC for most coating applications. Soft logic programs (programs to make a PC operate like a PLC) will still be introduced into the marketplace. SCADA systems will improve and begin to utilize the Internet for supervising and monitoring coating processes.

ASPECTS OF AN AUTOMATED COATING SYSTEM

A coating system (just like any industrial control system) is simply a group of interacting devices that perform specific actions in conjunction with one another in order to coat tablets. The series of actions that a manufacturer takes in transforming product A into product B is considered the coating system.

Webster's dictionary defines the term *automatic* as a "self-acting or self-regulating mechanism" (Merriam-Webster 1993). In a very broad sense, an automated coating system is a coating system with one or more self-regulating or self-actuating devices. Such a general definition leads to several problems since one person's perception of the term *automatic* is invariably different from another person's perception. This section will explore the various aspects of automated coating systems, how these aspects are defined, and the relationship between them and a coating process.

Many people, when they think about automated coating systems, immediately think of PCs, PLCs, and DCSs. However, there are more fundamental components that compose an automated coating system. In fact, the automated coating system can be broken into two basic elements that are required for automation: hardware and software.

Hardware

Hardware devices have physical attributes associated with them and are visible to the operator. Those associated with an automated coating system either perform actions during the coating process or monitor the status of the coating process. Examples of hardware devices used in coating applications include PLCs, OITs, temperature sensors, airflow sensors, valves mounted in the ductwork, and electric motors.

Hardware devices are typically used to automate control systems. An example of such a device was discussed in conjunction with a control relay. The control relay can be opened or closed, which then deactivates or activates an electric motor. A control relay can be activated when an operator presses a particular push button or when an alarm condition occurs during the coating process. In any event, the control relay can be wired to activate the electric motor automatically according to the requirements of the coating process.

A problem with using hardware devices to automate processes is that they have rigid control characteristics. Hardware devices are typically designed for a single purpose, and flexibility is often limited. For example, the control relay has only two states: open and closed. In order to design more complex automation into the system, additional control relays are required. Hardware devices typically can be classified in two categories: inputs and outputs.

Inputs

Inputs provide information about the status of the coating system. Inputs are the eyes and ears of the coating system. They provide feedback to operators so that they can determine if there is a problem with a particular device or how a coating variable (such as airflow or temperature) is reacting during the process. The selection of input devices for a coating system should be based on the process requirements for producing a consistent product. Typical inputs for coating applications include airflow sensors, pressure sensors, temperature sensors, and solution flow-rate monitors.

Outputs

Outputs are components that perform actions during the coating process. Outputs are the arms and legs of the coating system. They manipulate the critical process variables of the coating system. The selection of outputs designed into a coating system should again be based on the process requirements for producing a consistent product. Typical outputs for coating systems include electric motors, blowers to create the airflow through the machine, and valves located in ductwork to control the airflow.

Software

The other element in automated coating systems is software. Microsoft Windows® is an example of a software product. Software is, in its most fundamental form, a series of binary digits (ones and zeros known as bits) that cannot be easily interpreted by humans without the help of a computer. Computers can process several million bits in a single second, whereas humans can not comprehend information at these speeds. Therefore, unlike hardware products, software products do not have any physical attributes that can be associated with them. In other words, software cannot be heard, seen, or felt by humans.

The development of software products has been a driving force in the automation of coating processes. There are several inherent advantages to using software for automating coating processes, including flexibility, reliability, usability, and in validating automatic sequences. These advantages are described in more detail in the following sections.

Flexibility

Since software products do not have any physical attributes associated with them, there are, virtually, an unlimited number of functions that software can perform. This is in sharp contrast to hardware devices, such as a control relay, that may be specially designed for a single purpose and require additional devices in order to increase functionality. Software products, because they are flexible, can increase functionality as long as the computer storing the software (such as a hard disk drive) is large enough and can operate the software product in an effective manner. A major challenge in PC design today is to develop computers that have enough capacity and speed to operate software products more efficiently.

Reliability

A significant benefit of software products is that they do not deteriorate. For example, a software product that was installed on a computer 10 years ago is still the same software product today. Obviously, there could be problems, but they most likely would deal with the failure of the hardware components (a bad storage device or a failed processor). Since software products do not deteriorate or degrade, they are extremely reliable in that, given a certain set of input conditions, a predefined set of output actions will occur. This is an extremely beneficial property of software products and is one of the reasons why software has become such a driving force in any type of automation.

Usability

Software products are very good at performing a sequence of events. For example, a typical spray system first activates the nozzle air, then the cylinder air, and finally the pump (if recirculation is not used). The automation of this sequence can occur with an operator familiar with the sequence pressing three push buttons in succession. With software, the entire sequence can be programmed into a PLC; the operator would only need to press a single push button to initiate the spraying sequence. If this sequence requires modification (such as adding another pump to the spray system or changing the timed interval between steps), the software program could be modified to allow for these changes. Modifications would be much more difficult if software was not utilized as part of the automation process.

When a PLC, DCS, or PC is used to automate the coating process, most of the electrical devices are controlled directly through them. Therefore, the coating system can display multiple process variables on a single process screen on the OIT. For coating systems not using a PLC, DCS, or a PC, the process variables would have to be displayed using independent digital displays. For complex coating systems with more than 10 process variables that need to be monitored, independent displays would be cumbersome to monitor and control.

Validating Automated Sequences

The U.S. Food and Drug Administration (FDA) stated in their *Guideline on General Principles of Process Validation* that the purpose of validation was to "provide a high degree of assurance that a specific process will consistently produce a product meeting its predetermined specifications and quality attributes" (FDA 1987a). Software products will consistently perform a predetermined set of actions given that a predefined set of input conditions exists. Therefore, software products provide a "high degree of assurance" that the automatic coating sequence is being performed correctly.

Since software products are reliable, flexible, easy to use, and crucial for validating processes, utilizing software for coating systems often decreases the time it takes to develop new coating processes. The amount of time it takes to run established coating processes is also reduced.

Problems with Software Products

There are, however, some definite disadvantages in designing software for an automated coating system. Maintenance of the software product is difficult since there are no physical attributes associated with software. An additional computer is often required to modify software products. This could make maintaining and supporting software products expensive. If there is a problem with the software associated with the coating system, the problem may cause unnecessary "downtime," if correct programming devices are not available to modify the software. Even if the correct programming devices are available, a software "expert" is often required to fix the problem. Having a second or third party responsible for maintaining and supporting software products is both time-consuming and costly.

Software is extremely susceptible to poor programming. Without proper software development procedures and strict software design methodologies, software quality will vary with individual programmers. Since there are no physical limitations to software, an infinite amount of programs can be written to accomplish a single goal. Therefore, the quality of software is directly dependent on the programmer's ability to design quality into the software product. There have been several instances in every industry where computer glitches or malfunctions have caused tremendous problems, which are extremely expensive, and in some cases potentially dangerous (Weiner 1993). The third section of this chapter will discuss how quality is built into software products.

The Interaction Between Hardware and Software

Most coating systems today are composed of a mixture of hardware devices and software products. In fact, most devices incorporated into a system have both hardware and software. Temperature controllers, alternating current variable frequency drives, and digital displays have both hardware and software associated with them. Since there are both hardware devices and software products associated with coating systems, the following question may arise: Is software more important than hardware or vice versa?

The answer is that both hardware and software are equally important. If the quality of a coating system could be quantified, the entire system's quality would only be as high as the lowest value of quality between the hardware devices and software products. The reason for this relationship lies in the fact that software only improves the functionality of hardware devices. Software does not, and cannot, affect the accuracy, durability, or reliability of hardware devices.

Defining the Elements of an Automated Coating System

Every coating system can be divided into hardware devices and software products. Hardware can be classified as input devices and output devices. A drawing that can identify all aspects of a coating system is called the piping and instrument drawing (P&ID). An example of a P&ID is shown in Appendix B. This drawing identifies the general location of components within the control system and categorizes the components as either a software product, an input device, or an output device. Coating systems defined by a P&ID effectively "normalizes" the system into its most basic elements.

A "normalized" control system is independent of the type of controlling/monitoring devices (PLCs, DCSs, or PCs) that actually control the system. The P&ID "normalizes" the control system into its most fundamental elements by exactly defining the control devices that are part of the coating system. Both the software designers of the coating system and the process scientists who operate the system can understand the basic aspects of the automated functions. From the process scientist's perspective, this is an important concept. By "normalizing" the control system before design work begins, a process engineer can adequately define the requirements of the process without becoming an expert in software systems.

An accurate P&ID should also indicate the critical control loops that will be a part of the coating system. A control loop is, in its simplest form, a process variable that is controlled by an output device. Complex control loops can be designed with several process variables and output devices that are interrelated; however, the scope of this section will focus on the simplest form of control loops.

Typical control loops in coating processes include an input device, an output device, and a controlling device. The input device provides a measure of the process variable that needs to be controlled (such as airflow entering the tablet coater). The output device can be manipulated to control the value of a process variable (such as the position of a damper in the ductwork). The controlling device interprets the signal from the input device and modifies the output device to achieve the desired process variable.

Control loops are typically defined as either open or closed loops. An open loop is a control loop where the output device is controlled by the operator of the coating system (i.e., the operator is the controlling device). A closed loop is a control loop where the operator does not manipulate the controlling device. The operator is only responsible for entering a setpoint into the controlling device. The controlling device will manipulate the output device until the process variable is equal to the setpoint. Closed loop devices are very common in the coating industry. The most common closed loop controlling device is a temperature controller that controls the temperature of air entering the tablet coater.

There are several types of closed loop control philosophies including PID controllers and fuzzy logic controllers. Both types of control philosophies have been used in coating applications. The background behind these philosophies is beyond the scope of this chapter.

An example of a closed control loop, inlet air temperature control, is shown in Appendix B on the P&ID drawing. The tags *TE2300* and *TT2300* (located near the pan unit) represent a temperature sensor that is monitoring the temperature of airflow entering the coating pan. (TE2300 represents the temperature sensor—an RTD [resistance thermal detector]; TT2300 represents a transmitter that converts the signal from the RTD into a 4–20 mA current reading for the controlling device). After the tag *TT2300,* the tag *AI* indicates that it is an analog input and that it is connected to the controlling device (PLC, DCS, or stand-alone controller). The term *analog* indicates that the input control device gives a variable representation of the inlet air temperature. The loop then continues on to *TIC2300,* which represents the control loop inside the controlling device. After the tag *TIC2300,* the control loop moves onto three tags: *AO, IPT2300,* and *PV2300.* The tag *PV2300* represents the steam valve that regulates the amount of steam entering the heating coils. (The jagged lines represent the heating coils.) The tag *IPT2300* converts the electrical signal from the controlling device into a pneumatic signal that controls the steam valve. (IPT refers to an I-P transducer where I represents the electrical current and P represents the pneumatic control pressure.) The tag *AO* indicates that *FCV230* is an analog output connected to the controlling device. With these seven tags, the P&ID has depicted a tremendous amount of information relating to the coating system. The P&ID is very important because it can depict a vast amount of information for both the process scientist and the software engineer in an easy and efficient manner.

Quality Means and Measuring That Quality

It has been stated that the overall quality of a coating system is only as good as both the hardware and software associated with that system. What are the means to achieve high quality? How does one measure the quality associated with these two aspects of the coating system?

Hardware Quality

When thinking about the concept of "quality," people often think in terms of hardware devices. This probably occurs because "quality" is easier to understand when applied to hardware devices. When measuring the quality of a hardware device, a person can physically test the attributes of that device, the accuracy of its output, and the environmental conditions in which it can maintain accurate outputs. Statistical process control (SPC) is primarily based on the measure of quality applied to hardware products.

Components that make up the hardware device must be of a "high quality" in order for the entire system to be considered "high quality." If human intervention is required to assemble the device, the human must be properly trained and perform several quality reviews in order to achieve high quality standards.

Software Quality

Since software products do not have any physical attributes, one cannot easily judge the quality of software just by operating the coating system for which the software was designed. The FDA recognized this fact and wrote a technical report entitled *Software Development Activities* (FDA 1987b). This report was written to establish general guidelines on how to approach software development in the pharmaceutical industry. The following is an excerpt from this technical report:

> Software can be very complex. A process control program may consist of thousands of lines of code with numerous paths and branches from instruction to instruction. Nevertheless, the logic of the program must be understood by those responsible for writing, testing, and maintaining it. Because of this complexity, and the fact that computer systems have progressed from performing only basic arithmetic calculations to life dependent and life sustaining systems, computer software development has evolved into a recognized science of established procedures and disciplines. Among the basic teachings of this science are two premises:
>
> 1. Quality begins at the design level.
>
> 2. You cannot test quality into your software.
>
> In an attempt to assure that these premises are followed, all categories of software should go through an organized development process consisting of a series of distinct phases

which are collectively referred to as the SOFTWARE LIFE CYCLE. The existence of these identified phases allows reviews and tests to be conducted at distinct points during the development and installation of software. This in turn helps build quality into the software and provides a measure of that quality.

This excerpt implies that software should be developed consistently, without regard to the type of coating process for which the software is being developed. For example, the functionality of software developed for a film coater would be significantly different from the functionality of software developed for a sugar coater. However, the design and development of both software systems should be consistent with each other. The FDA states that the method by which quality can be built into software systems is to develop the software under a software life cycle, which will be discussed in the next section.

If the means of ensuring a high quality software system is to design the software under a software life cycle, then what measures the quality of that software product? Measuring quality can encompass several levels. From the process scientist's standpoint, measuring the quality of software modules includes testing the correctness, reliability, speed, and usability of the software system. From a designer's standpoint, measuring software quality would also include testing the reusability, maintainability, and verification of the software modules used to develop the software system.

The FDA formalized a method in which the quality of software systems can be measured. This formalized method was called validation. Enforcement of the validation of software products began in the late 1980s in an effort to measure the quality of software products designed for pharmaceutical applications. Validation includes formalized tests that measure the quality of all aspects of the coating system, including the following (FDA 1987a):

- Tests that prove the devices that comprise the coating system are installed correctly (installation qualification [IQ]).

- Tests that prove the automated sequences defined by the software system are repeatable (operational qualification [OQ]).

- Tests that prove the products produced by the coating system are consistent from batch to batch (performance qualification [PQ]).

It is virtually impossible to test every path and line of code within a complex software product. This is just one of the reasons why the FDA stated that quality cannot be tested into a software system (premise #2). The presence of formalized test procedures does not guarantee a high-quality software system. Instead, formalized test procedures only provide a measure of software quality.

Why Is Software Quality Important?

Why is purchasing a high-quality software system desirable? There are three reasons for a high-quality software system:

1. A software product of high quality minimizes the inherent problems of software described earlier in this section. Software products that were designed under the same software life cycle would produce products of the same "high quality" even if the designs of the software products were developed by different people. Since the software would be developed in a similar manner from designer to designer, support for the software system would not be limited to the original programmer.

2. The time it takes to check out and validate a software product of high quality would be less than the time it would take to check out and validate a lower quality system.

3. A high-quality software product should give the process scientist a higher degree of confidence in trusting expensive production batches to something they cannot see, hear, or touch.

THE SOFTWARE LIFE CYCLE

The software life cycle identifies distinct phases that software products undergo before a coating system is completed. It is important to realize that there are several different kinds of software life cycles that can ensure a high-quality software product. A simplified version of a typical software life cycle is displayed in Figure 5.2 (IEEE 1991).

Phase 1: The Definitions Phase

The definitions phase is the first phase of the software life cycle. During this phase, the designers of the system generate several documents that define WHAT actions the coating system is to perform.

Figure 5.2. Software life cycle.

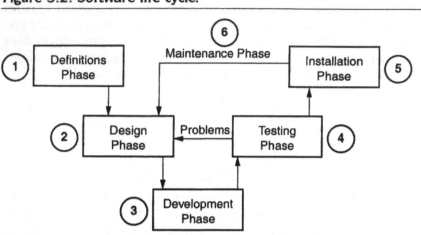

These documents are listed below and described later within this section:

- P&ID.

- Standard Operating Procedures (SOPs).

- The Functional Requirements Documents (FRDs).

The P&ID was discussed earlier in this chapter. The main purpose of the P&ID is to define the hardware devices (both inputs and outputs) that are to be part of the coating system. The P&ID also defines the control loops that will be associated with the coating system.

Software SOPs define the details of designing, developing, and maintaining the software. These procedures ensure internal consistency throughout the software program, as well as external consistency with other software products. Software SOPs are typically developed and maintained by the designer of the software system. The purchaser of the system should audit the designer before purchasing a software system. This audit should judge the completeness and accuracy of the software SOPs maintained by the designer of the coating system.

The purpose of the FRD is to define the requirements for a specific automated coating process. Both the vendor and the purchaser must be responsible for the FRD. In some cases, the FRD is nothing more than the sales proposal from the designer and the purchase order from the purchaser. The FRD includes more specific information than what might be specified on the P&ID. For example, the P&ID might state that a motor is required to rotate the pan, but the FRD would specify a 30 hp motor powered from a 460 V alternating current source. Typically, the FRD is the legal document (or documents) that binds the designer in supplying a coating system to the purchaser for an established price.

Phase 2: The Design Phase

The design phase is the phase at which the designer determines HOW the software product will perform the functions defined in the definitions phase (IEEE 1991). The design phase can be divided into two aspects: developing software from an established baseline system and writing the software functional specifications.

Development from an Established Baseline System

The term *baseline* can have different meanings to different people. In terms of software development, a baseline is a starting point for design. Baseline systems are composed of documentation and source code that are similar to the system defined in the definitions phase. Baselines that are used as starting points for development are often called functional baselines (Berlack 1992). If bugs (i.e., errors) or inconsistencies are found with a functional baseline during the development of the software product, a procedure must be in place to correct the functional baseline before another design engineer uses it again. By doing this, the functional baseline will become error free over time. If the starting point used for software development is error free, there will be a reduced amount of errors in the final software product. Purchasers should ensure that the designer of the software product has established baselines that will be used as the starting point when developing new automated coating systems.

The first step in the design phase is to read and understand the documents created in the definitions phase and select the correct functional baseline.

Writing Software Functional Specifications

After the functional baseline has been selected, the design engineer will write software functional specifications for the proposed software system. Software functional specifications define HOW the software system will meet the requirements that were established during the definitions phase. The software functional specifications will describe all of the details of the software system (FDA 1987b):

- The activities, operations, and processes that are being controlled, monitored, or reported by the computer system.

- The software used during the development process.

- The files that are accessed/created by the system.

- Limits and parameters to be measured by the system.

- How the process parameters are controlled/monitored for a specific process.

- All error and alarm messages, their cause, and the corrective actions that must be taken should they occur.

- All interfaces and communication modules required between the processor and the external world.

- The security measures that are to be followed to protect the program from both accidental and intentional abuse.

Functional baseline documentation should be used as the starting point for the software functional specifications generated during the design phase.

The most important thing to remember about the design phase is that there are no devices being programmed at this point. The output of the design phase should only be the documentation that defines how the system will satisfy the requirements specified in the definitions phase. The FDA stated that quality must be built into the software during the design phase (premise #1). If documentation is generated that defines how the software system is to perform BEFORE actual programming occurs, quality will be built into the software system.

The purchaser of the software system should always review and approve the proposed software functional specifications before programming begins. The purchaser will then have a better idea of what to expect when the system is completed, and the designer will be able to identify future problem areas that could arise while programming the devices.

Phase 3: The Development Phase

The development phase is the phase at which the design engineer enters the software code into the programmable device. The software code is generated from the approved software functional specifications generated in the design phase. Another important objective of the development phase is to perform informal testing of the software system while the software is being developed. In fact, most of the testing performed on the software system should be performed during the development phase. Typically, the design engineers who do the programming are the same people who perform the informal testing.

Phase 4: The Testing Phase

The testing phase can be separated into two elements: informal and formal testing. It was mentioned that informal testing is performed during the development phase. However, the person responsible for developing the software is also responsible for the software during the testing phase. As a result, the same person developing the software product typically performs informal testing during this phase.

A method useful for the designer to test the software informally during this phase is to write simulation programs. Simulation programs are routines that simulate the coating process without the hardware devices connected to the software system. Depending on the complexity of the simulated system, these routines can be written by the actual programmer or purchased from a third party. Either way, simulating the system before formal testing begins is an invaluable tool in eliminating the errors found in software products during this phase.

Another qualified person who ensures that the program conforms to the documents defined in the definitions and design phases usually performs formal testing. Validation procedures are written to test the software system formally. These procedures were briefly described earlier under "Software Quality." Validation engineers specialize in writing procedures that can be understood by auditors who may not be familiar with programming techniques.

While the software product is being formally and informally tested, problems will occur and should be expected. Logical, grammatical, and structural errors within the software functional specifications could cause the software system to be programmed correctly while not performing the requirements established during the definitions phase. Syntax errors could occur during the development phase—mistakes where the program does not match the software functional specifications. The documents approved in the design and definitions phases should be challenged, and the appropriate corrective action should be determined when problems arise. Both the designer and the purchaser of the software system should approve changes to the FRD or P&ID if they are necessary.

Phase 5: The Installation Phase

The installation phase is the phase at which the software product is installed and formally tested at the purchaser's site. Due to the complexity of some coating systems, not all software can be tested at the designer's production facility. In these cases, software products that could not be formally tested during the testing phase must be tested at the purchaser's facility. Also, during the installation phase, hardware components that are part of the system are calibrated by the vendor of the coating system.

Testing is required to verify that the software and hardware devices work correctly. In addition, process testing is required to ensure that the coating system can produce acceptable product. If it is proven that the coating system cannot produce consistent products, the design documents generated in the definitions phase (FRD and P&ID) should be modified, with new software products developed as necessary. Obviously, both the vendor and the customer will want the changes to the system as fast as possible; however, it is imperative that the development of additional software products still follow the same software life cycle.

Phase 6: Maintenance Phase

After the entire coating system (both hardware and software) has been verified, and production testing has yielded a successful and consistent product, the software product enters the maintenance phase. The maintenance phase is the phase at which the function of the software product is maintained and modified and/or enhanced as necessary. The vendor is typically responsible for ensuring that the integrity of the completed software products will be maintained. Modifications to the completed system must be performed according to the software life cycle.

In Figure 5.2, the arrow from the maintenance phase is drawn to the design phase. One might ask, "Shouldn't the arrow be drawn to the very beginning of the software life cycle?" The answer to this question lies in understanding how support for the software product will be performed. After the coating system is completed, a complete set of documentation (the P&ID, the FRD, and the software functional specifications) will have been generated. In supporting the existing coating system, the designer will use these documents as the starting point if changes to the software are required. The arrow from the maintenance phase to the design phase implies that the existing documentation can be used as the "starting point" for supporting the existing system. The arrow from the definitions phase to the design phase implies that the SOPs required throughout the development of the software product are still applicable in supporting the existing system.

Determining the Quality of Software Systems

Much has been discussed about how to design a software system that is of "high quality." But how can a purchaser of a coating system, who may not develop software as a profession, determine the quality of software products produced by a vendor? The primary indicator of software quality is in the amount of design specifications produced *before* the software product is developed. Notice that design specifications, not validation test procedures, are the primary indicator of high software quality.

Consider two vendors who have developed software for coating applications. The first vendor, in an attempt to establish high software quality, presents binders of test procedures proving that their system has been completely tested. The second vendor presents the FRD, P&ID, and software functional specifications of the software system before it was designed. Furthermore, the second vendor backs these documents up with validation test procedures that prove the performance of their coating system. Since quality cannot be tested into software, the purchaser of the coating system should conclude that the second vendor has higher software quality than the first vendor. The first vendor did, in fact, perform several tests on their system, but they never proved that these procedures adequately tested what the system was originally designed to perform.

OPERATOR INTERACTION

While the software life cycle provides a means to develop high-quality software, it does not provide insight for the designer to develop a coating system that is user-friendly. Developing a system that is user-friendly is more of an art than a science. What makes developing a user-friendly system difficult is that every user who interacts with the system has a different definition of the term. A system is typically considered user-friendly if the users interacting with the control system understand and interpret correctly the information presented to them on the display screen. This section will discuss some guidelines that, if followed, should provide both purchasers and designers of coating systems some guidelines of how modern day automated coating systems should be designed.

Graphical User Interfaces

Most automated coating systems have some type of OIT that presents information to the operator. These displays can sometimes graphically represent the coating system on the display screen. A graphical representation of a coating system helps the user understand how the different components of the coating system interact with one another. The graphical user interface (GUI) is essential for individuals who are not familiar with coating processes to understand the processes that are being controlled and monitored. However, for users who are familiar with coating systems, graphical screens may be redundant and unnecessary. Often, experienced users want to see only the information about the process and can interpret the information without the coating system being represented graphically.

From the developer's perspective, graphical systems seem to impress prospective customers. Many purchasers of coating equipment incorrectly relate high-quality graphics with a high-quality software product. However, as discussed previously, developing software under a software life cycle is the method in which quality is built into software products—not impressive graphics.

There are several questions that should be answered by the users of the coating system before the amount of graphical representations can be determined for a system.

- How many users will be interacting with the coating system?

- What functions do these users perform when interacting with the coating system?

- Where will these users be interfacing with the coating system? Many times, there can be multiple display terminals associated with one system.

- How familiar are the users with the system?

- How familiar will new users be with the system?

Consistency in graphical interfaces also plays a major role in developing a coating system that is easy to understand. Most coating systems on the market today will have consistent graphical interface standards that make a coating system more user-friendly. These standards are described below:

- Modifiable process parameters should be identified by a method unique to the rest of the coating system. A user should be able to look at the OIT and immediately determine which process parameters are modifiable and which parameters are only used for monitoring the coating process. Some vendors make modifiable parameters a different color, while others draw boxes around the modifiable parameters.

- Functions that need to be addressed by the user should be highlighted or identified in a consistent manner. There is nothing more frustrating to a user than not being able to perform one function because another function must be performed first. By identifying on the OIT the functions that require immediate attention, the user should be able to identify more efficiently the next function that needs to be performed.

- Similarly, functions that are disabled should be identified on the OIT. A common method of indicating a disabled function is by "dimming out" the text on a push button. For example, if all of the enabled functions on the screen have text that is black on the associated push button, then the disabled functions would have text that is light gray.

- Users should look at a process screen and be able to determine immediately the different function keys (or push buttons) that are available from that screen. By consistently locating the function keys on similar areas of each screen, the operator can quickly determine the available functions associated with the coating process.

- A method of indicating which functions are activated or deactivated should be consistent throughout the control system. An operator, from a single glance at the OIT, should be able to determine which devices are currently enabled and which are disabled.

A disadvantage of GUIs has been their high cost. A device that supports a GUI is typically more expensive than a device that does not support graphics. Purchasers of coating systems must always ask the question, "How important is it to purchase a coating system that is easy to understand?" GUIs are becoming more common throughout the coating industry, especially since the cost for these devices has been decreasing while device functionality has been increasing.

Software Modes

Software modes are areas of the software system that perform specific functions. Software modes can be separated into two major classifications: process control modes and supervisory control modes. Process control modes are modes that directly control the input and output devices associated with the automated coating system. Supervisory control modes are modes that configure how the system performs or manipulates/displays historical process information. The following sections will discuss how automated coating systems are commonly separated for batch and continuous coating applications.

Process Control Modes—Manual Mode

Operating a coating process in the manual mode is probably the most common mode associated with an automated coating system. The manual mode is most often designed for the development of new processes or for processes that need constant operator attention. Due to its name, confusion arises in defining the amount of automation associated with a manual mode. A well-structured manual mode will only allow the user to control the major functions of the coating system manually, while automatically controlling the minor functions. Examples of what would be considered major and minor functions of the coating process are given below.

Consider again the temperature control of process airflow entering the tablet coater. Process air temperature is typically controlled automatically (through a controlling device) to a certain setpoint defined by the operator. Having an operator continually adjust a heating device manually is probably not repeatable and certainly not practical. In this example, adjusting the heating device would be considered a *minor function*. However, the manual mode must provide an operator to activate the controlling device manually such that temperature control will begin. Determining whether or not the device controlling temperature should be activated would be considered a *major function*. In many applications, the ability to initiate a control loop manually and let the software system automatically attain a set value is very common in the manual mode.

Another example of automatic control in the manual mode is a simple spray system that does not utilize a recirculation return line for excess solution. In this example, activating the airflow through the guns before the solution pump is activated will eliminate drops of solution falling on the product bed. Typically, a well-structured manual mode would not let the operator press two buttons in close succession in order to initiate spray. Instead, one button could be offered to the operator that would automatically activate the spray sequence without dripping solution onto the product. Activating the airflow through the guns and activating the solution pump would be considered *minor functions.* The entire spray system would be considered a *major function.*

Purchasers of coating systems need to think about which processes should be manually controlled and which should be controlled without operator intervention.

Process Control Modes—The Automatic Mode

An automatic mode is most commonly utilized for processes that are well defined and repeatable. For many customers, the automatic mode is the main reason for purchasing an automated coating system. A well-structured automatic mode should perform the following tasks: execute a predefined recipe (recipes are defined later in this chapter), compensate for process variables that are out of specification, and perform a set of predefined actions if hardware devices fail during the process.

If problems occur during operation, the coating system typically places the process in a state that suspends execution of the process while protecting the product. This state is often called a *hold state.* For coating applications, typical hold states consist of deactivating the heat and jogging the pan at predefined intervals. The operator should also have the capability of placing the process in a hold state in case problems with the process occur that cannot be detected by the automated coating system.

In some cases, problems occur such that the only recourse is to abort the automatic process and perform a controlled shutdown of the system. The coating system needs to ensure only operators with the proper security clearance can abort a process.

A well-structured automatic mode would allow an operator to monitor the predefined recipe while the automatic process is running. This is important from the standpoint of understanding the actions that will occur next in the process and perhaps compensate for potential problems.

Process Control Modes—The Maintenance Mode

A maintenance mode is most commonly utilized during equipment calibration and maintenance. A well-structured maintenance mode should allow the user to control and monitor input and output devices without the typical process interlocks and restrictions associated with manual and automatic modes. For example, consider the simple spray system described above. A well-structured manual mode would combine activating the air through the spray gun and the solution pump into a single push button. With the typical process interlocks removed, a well-structured maintenance mode should allow both the air through the guns and the solution pump to be activated independently from each other.

Input and output devices should be grouped together and displayed in such a manner that as many devices as practically possible are visible. Safety interlocks associated with the hardware devices should still be enabled throughout the maintenance mode. Output devices that could damage equipment or potentially harm people should be restricted from use in the maintenance mode.

Supervisory Control Modes—Define Recipes

In order to run any process in the automatic mode, there must be an established sequence of events for each automatic process. Every established sequence is made up of multiple process steps. For example, the sequence could include preheat, spray, and cool-down steps. Within each process step, there would be a set of instructions that would define exactly what actions the coating system is to perform. These sets of instructions typically include the following:

*   A definition of when major functions are activated and deactivated during the process. For example, temperature control would typically be deactivated during a cool-down step, whereas the solution pump would be activated during a spray step.

*   The ability to modify setpoints associated with major functions.

*   The ability to modify alarm values that are associated with major functions. Temperature control would typically have an "overtemperature alarm" as part of the automatic sequence. If the temperature exceeds this alarm limit, the coating system would know that the process should be placed in a hold state until the alarm condition is cleared.

These instruction sets are generally referred to as the process recipe for the automatic mode. There are two basic methods of managing automated coating recipes: the "grid method" and the "object-oriented method."

The Grid Method. The grid method received its name from its appearance on the OIT. The grid method displays the process parameters associated with the automatic sequence in a spreadsheet-like format. See Appendix C for an example of a process screen that utilizes the grid method. The top of the grid represents the major functions associated with the automatic sequence. The left side of the grid lists the process steps associated with the sequence. To define a recipe, the operator would have to fill out the grid by entering values for the appropriate setpoints. The grid method is quite simple to use; however, its major drawback is that it is not very flexible. For example, the changing of setpoint values within the established grid is simple. However, if the user has an application where an additional setpoint or process step is required, software modifications to the coating system are often needed. The grid method lends itself well to processes that are well defined or a coating system that runs only one process. If the coating machine runs a variety of products that have different recipe sequences, the grid system may not be the ideal recipe management system.

The Object-Oriented Method. The object-oriented method is relatively new to the tablet coating industry. It has grown a great deal lately due to the advances in PCs and operating systems for those computers. The object-oriented method allows the user to build a recipe with virtually an unlimited number of process steps. This is usually accomplished by breaking the coating process into its major functions. An example of an object-oriented recipe management system is shown in Appendix D. The process steps are still displayed on the left side of the screen. However, the major functions of the coating system are now identified with graphical icons. These icons can be easily added to and/or removed from any process step. For example, according to Appendix D, the first process step is made up of an airflow icon and a temperature icon. This would indicate that the first step of the coating sequence would consist of the activation of airflow through the pan and controlling the temperature of that air.

With the object-oriented recipe, additional process steps can be easily added to and removed from the coating sequence. The user would be able to access the process setpoints and alarm conditions associated with the icons and establish the system's operating parameters. This method is in sharp contrast to the grid method, where adding/removing major functions to the automatic sequence would require additional software modifications. The only drawback to object-oriented recipes is that they are often more difficult to understand than a static grid recipe.

As mentioned previously, the grid method is simple but inflexible. The object-oriented method is flexible but more difficult to comprehend. Recipe management systems of the future should have the ability to transfer information between a grid recipe and an object-oriented recipe with minimal effort. By combining both types of recipe management systems, users can enjoy the benefits of both methods.

Finally, future technology should also allow for process information to be saved in the manual mode and imported directly into the recipe management system as complete steps to the automatic sequence. This will allow the operator to develop processes in the manual mode and save those process parameters that were used in developing a consistent coating process. These process parameters can then be retrieved from the system and entered as a complete process step in the automatic sequence, thus reducing the amount of time it takes to transfer products being developed in the manual mode into an automatic sequence that can be run in the automatic mode.

Supervisory Control Modes—Electronic Batch Reports

Electronic batch reports are primarily associated with coating systems that utilize personal computers as the OIT. Typically, handwritten batch reports are generated by operators to document the process parameters of the coating process. These reports would then be submitted to regulatory organizations (e.g., the FDA) to prove that the product was produced correctly and that it can be distributed to the public. Electronic batch reports simply record the process parameters electronically instead of on paper. The electronic reports can be printed out and submitted to regulatory agencies in much the same way as handwritten batch reports.

A significant advantage to electronic batch reports is the accessibility to anyone in the plant who is interested in monitoring processes, trending process variables, or verifying that the process was performed correctly. Electronic batch reports also would reduce the amount of paper that would be required at a manufacturing facility.

The major disadvantage to electronic batch reports is the susceptibility of electronic files stored in a poorly secured network. Just as handwritten batch reports are stored in a secured location, electronic batch reports should also be stored in a secure location on a plantwide network. Electronic batch reports could be easily modified if they are not stored in a secure location.

If electronic batch reports are a requirement of the coating system, a well-structured supervisory control mode must be in place to control users who want to display, print, modify, or delete these reports. For obvious reasons, this mode must be protected by a sophisticated security system. Future technologies should allow the operator to customize electronic batch reports differently for different coating processes.

Supervisory Control Modes—Historical Trending

Another important source of information that is available with PCs is historical trending. Historical trending is nothing more than graphing the process parameters of a coating process that has already occurred. A historical trending mode should be made available for operators to generate trends of process parameters of completed coating processes. A well-structured historical trending mode would link the historical trends with the electronic batch reports.

Integration of Software Modes

With modern automated coating systems, process control modes and supervisory control modes will be accessible from the same screen on the OIT. The speed associated with OITs today makes it possible for an operator to have several screens that are associated with independent software modes active at the same time. For example, while a coating process is being operated in the manual mode, an operator will be able to access electronic batch reports in order to compare the current process with a historical process.

Supervisory Stations

Pharmaceutical manufacturing companies are beginning to realize the need to monitor, control, and service coating systems from multiple locations. Multiple locations may be defined as the process room next to the coating equipment or a location halfway around the world. This type of remote monitoring/control is commonly referred to as supervisory control.

Supervisory control is the next step in coating system functionality. It typically implies that the user has the ability to monitor, manipulate, and control a process from a remote location. Supervisory control stations are also being used as information hubs. Recipe generation, historical data retrieval/trending, and batch reporting from multiple coating machines are considered base requirements for supervisory control stations.

Another benefit of supervisory stations is multitasking. A well-designed supervisory station should allow the user to monitor multiple coating processes simultaneously. It should also be designed to allow for things such as recipe creation and retrieving electronic batch reports while running one or more machines at the same time.

BATCH CONTROL STANDARDIZATION

A term that is frequently heard throughout the pharmaceutical coating industry is *S88*—a recent standard that the International Society for Measurement and Control (ISA) has developed to define how batch control should be approached. The real name for this standard is ANSI/ISA-S88.01. This standard was developed to provide process engineers with standard models and terminology for batch manufacturing systems. It was also developed to provide a common forum for defining good practices for the design, operation, and control of batch manufacturing systems. It is important to point out that this standard applies to all batch systems regardless of the automation level.

What are the benefits of S88? According to the ISA (1995), following this standard will provide several benefits, some of which are listed below.

- It will decrease the user's time to take a proposed design from the drawing board to production.

- It will enable various vendors to provide the correct tools for implementing batch control.

- It will enable the user to better understand and identify its needs.

- It will make recipe development straightforward enough so that any necessary changes can be implemented without the need for a software engineer.

- It will help reduce the cost of automating batch processes.

On the other hand, S88 is not intended to suggest that there is only *one* method for controlling batch processing. It does not suggest that users should abandon all of their current practices nor does it try to restrict the creative development of batch systems.

The S88 standard can be divided into three sections: batch processes and equipment, batch control concepts, and batch control activities and functions. The following sections will briefly overview S88 in order to provide a flavor of the actual standard.

Batch Processes and Equipment

According to the ISA (1995), "A process can be defined as a sequence of chemical, physical or biological activities for the conversion, transport or storage of material or energy." A process can be classified as continuous, finite quantities of parts, or finite quantities of material (batches). The focus of S88 is only for batch processes.

This section is intended to help characterize processes into a standard format that can be used to help design, implement, operate, and understand a particular batch process. S88 defines all batch processes by a model. This model breaks down every process into four distinct subdivisions: process, process stages, process operations, and process actions.

Process

The process refers to the most fundamental function of the batch process (e.g., tablet coating).

Process Stage

The process stage can divide the tablet coating process into individual parts or steps (e.g., preheat, spray, and cool-down).

Process Operations

The process stage can be subdivided further into process operations. For a tablet coating process, the preheat product bed step could be broken down into heat the tablet bed at an inlet temperature of 65°C at an airflow of 2000 CFM, while maintaining a negative static pressure inside of the coating pan of −1.5 in. of water until the exhaust temperature reaches 40°C.

Process Actions

Process operations can be further subdivided into process actions. A process action describes the minor process actions required to perform the process on a machine level. For example, control inlet temperature to 65°C, open the main steam heat valve, activate the inlet blower, and so on.

Physical Structures

The physical structures section is intended to help characterize physical structures, or equipment layout and configuration, into a standard format. Three basic physical structures are included in S88: single path, multiple path, and network. By defining a process in this way, it is very easy to breakdown a system into its functional components. Once the functional components are defined, it is easier to get a process design from paper to product.

Single Path

A single path refers to a process arrangement where the batch travels through the structure in a single, predefined path. The single path structure has only one entry point and one exit point. The path structure may be on one machine or multiple machines. Several batches may be in progress at the same time.

Multiple Path

A multiple path refers to a process arrangement where multiple, single-path structures are running in parallel and no product is exchanged between paths. For example, a common raw material bin could feed several machines doing the same process, and the finished product from each machine would empty into a common bin.

Network Path

A network path is the most complex process arrangement and can be described as a batch sequence that could be routed through several different configurations. A path may be fixed for a particular batch or recipe or it can be variable.

Batch Control Concepts

The batch control concepts section of S88 discusses the batch control concepts needed for the process elements that were discussed previously and defines a consistent way of achieving these concepts in a batch manufacturing plant. It also addresses the concept of recipes and what properties a recipe should possess. This discussion will focus on the recipe portion of S88.

According to the S88 standard, a recipe is defined as "the necessary set of information that uniquely defines the production requirements for a specific product" (ISA 1995). This standard characterizes all types of recipes as either general, site, master, or control recipes.

General Recipe

A general recipe is a global recipe that serves as a basis or starting point for all lower-level recipes. This recipe is created without specific knowledge of the process equipment that will be used to create the product. Furthermore, it is not specific to a particular piece of equipment or plant site. It could identify, for example, the raw materials needed and their quantity. An example of a general recipe would be the film-coating process for drug XYZ that is able to coat 1000 kg of tablets in 4 h. The general recipe is typically used for plant planning and investment decisions.

Site Recipe

The site recipe is a recipe that is specific to a particular site and is a combination of the general recipe and site-specific information. This recipe may or may not be created from the general recipe. It could include, for example, local language for operator interfaces and programming or local raw material differences. Like the general recipe, the site recipe is not specific to a particular piece of equipment.

Master Recipe

A master recipe is a more detailed representation of the process recipe and is usually derived from the site recipe. The master recipe could be created on its own, but this requires the creator to have a great deal of knowledge about the process and the product. Some characteristics of master recipes are as follows:

- Multiple master recipes can be created from one site recipe. Each master recipe covers only a part of the requirements stated in the site recipe.

- The master recipe has to be equipment specific in order to ensure that the equipment will be capable of performing the desired process correctly.

- The master recipe will usually include the formula data that are needed.

- The master recipe may be product specific.

- The master recipe is a required recipe level. Without the master recipe, a control recipe cannot be created.

- The master recipe can exist as an electronic entity programmed into the equipment control or as an identifiable set of written instructions.

Control Recipe

The control recipe starts as a copy of a specific version of the master recipe. This copy is modified for a specific batch or product. It has enough detail to describe every operation of the machine and should be flexible enough to account for changes over time.

A good recipe structure should define five sections: header, formula, equipment requirements, recipe procedure, and other information. The header section should include administrative information in the recipe (e.g., recipe name, machine name, product name, version number, originator, date issued, and approvals).

The formula section should include process inputs, parameters, and outputs. The process inputs can include information about the raw materials used or the material preparation. The process parameters should include all setpoints, step transitions, and conditional logic to describe the recipe. The process outputs should describe all of the process outputs of the system, including finished materials information and environmental impact.

The equipment requirements section should include approved equipment to be used with the process.

The recipe procedure or configuration section should allow the recipe creator to use any quantity or combination of the recipe procedural elements that are predefined for the system. These procedural elements are fixed and cannot be changed by the recipe creator.

S88 also addresses production information, or what information a system should be able to produce. Batch specific information may include the following:

- A copy of the control recipe that was actually run.

- Recipe data that was collected during the process run.

- The system should allow for operator comments.

- The system should record event data such as alarms and operator activities.

All recorded information that is specific to a particular batch is called a batch history. It is important to note that S88 states that the batch history must be stored in such a way that the data can be associated with a particular batch.

Batch Control Activities

The batch control activities section of S88 discusses various control functions for batch processing, manufacturing, and control tasks. It also covers recipe management, production planning, scheduling, production information management, process management, unit supervision, process control, and personnel and environmental protection. This section is intended to identify the functionality associated with batch control. The following discussion will focus on the various aspects of production information management.

The information gathered for a particular batch may be as important or even more important than the recipe itself. Being able to prove that a batch process did, in fact, perform as expected is very vital to many people and companies in the batch process industry. S88 covers various parts of production information, including storage guidelines, reliability of batch history entries, batch and material tracking, logging of information, handling late information entry, manipulating historical data, and elements of batch reports.

S88 refers to the creation of batch reports as process management logging. Process management logging should include information associated with initiating and routing the batch and the equipment-independent information associated with the batch:

- A copy of, or reference to, the master recipe.

- All process management events and control recipe information, including any changes that were made to the execution of the control recipe.

- All operator comments, narrative descriptions, and operator observations of the batch processing. This information entry should be capable of being recorded with the operator's identification.

All process data for the batch history should be collected and stored in a way that includes or gives simple access to the following information:

- Batch identification.

- Absolute time stamp (real time).

- Identification of procedural elements with which the data are associated.

- Time relative to the start or end of a batch or of the execution of a procedural element.

- Equipment-independent entry identification.

- The equipment being utilized.

S88 also gives guidelines to the content that should appear in any batch report. The batch report can be either electronic or on paper. Some possible elements of a batch report include the following:

- *Report header:* Information on the report type, batch or batches displayed in the report, descriptive text, and so on.

- *Single elements:* Data elements are displayed somewhere on the paper/screen.

- *Event lists:* Chronological lists of event-type entries with associated data, including alarms or operator interventions.

- *Merging of entries in event lists:* Entries with different tags and of different types may be merged into the same list. For example, the report could include process data from various aspects of the coating machine, such as inlet temperature or coating pan speed.

- *Selection of entries into lists:* The batch report creator should have the ability to select what will and will not appear in the batch report.

- *Trends:* A batch report can include process trending that will display the process information graphically with one or more process variables on the same time axis.

- *Time-series:* The batch report creator should have the ability to select the report "time-deadband"—the interval that the process data should be displayed in the report. This is sometimes referred to as the print interval.

Summary

S88 is intended to be used as a guide for designing, implementing, and operating all areas of batch control systems. It is not a set of rules or regulations that need to be followed to the letter. It is a very useful tool for allowing all aspects of a system to be defined in a common form, such that all parties involved with a particular process will fully understand the system required. Vendors of equipment cannot be expected to follow everything that S88 covers. It is important that the vendor be aware of S88 guidelines and that they can explain how S88 applies to their system.

REFERENCES

Berlack, H. R. 1992. *Software configuration management.* New York: John Wiley & Sons, Inc.

Considine, D. M. 1985. *Process instruments and controls handbook.* New York: McGraw-Hill Book Company.

EC. 1995. *Electromagnetic compatibility (EMC) directive (89/336/EEC) as modified by the CE marking directive (93/68/EEC) and directives 91/263/EEC and 92/31/EEC.* The Council of the European Communities.

FDA. 1987a. *Guideline on general principles of process validation.* Rockville, Md., USA: Food and Drug Administration.

FDA. 1987b. *Software development activities.* Rockville, Md., USA: Food and Drug Administration.

IEEE. 1991. Standard for developing software life cycle processes, No. 1074-1991. Institute of Electrical and Electronics Engineers.

ISA. 1995. *Batch control—Part 1: Models and terminology,* ANSI/ISA-288.01. The International Society for Measurement and Control.

Merriam-Webster. 1993. *Merriam Webster Collegiate Dictionary,* 10th ed. Springfield, Mass., USA: Merriam-Webster, Incorporated.

Nillson, J. W. 1987. *Electric circuits.* Reading, Mass., USA: Addison Wesley Publishing Company, Inc.

Wiener, L. R. 1993. *Digital woes.* Reading, Mass., USA: Addison-Wesley Publishing Company.

Wilbanks, W. G. 1996. 50 years of progress in measuring & controlling industrial processes. *Measurement & Control Magazine* (February).

Appendix A: A Schematic of a Section of an Automated Coating System That Utilizes Electro-mechanical Devices

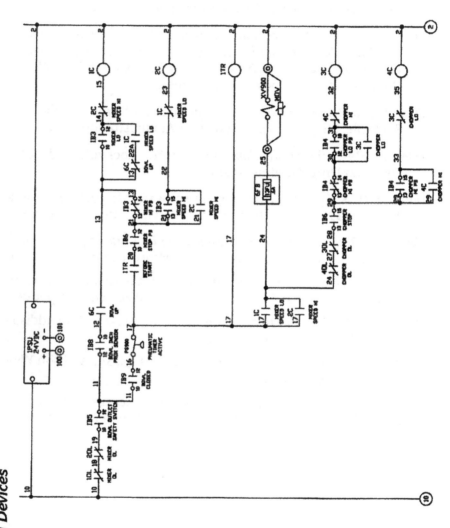

Appendix B: A P&ID Showing the Basic Elements of a Control System

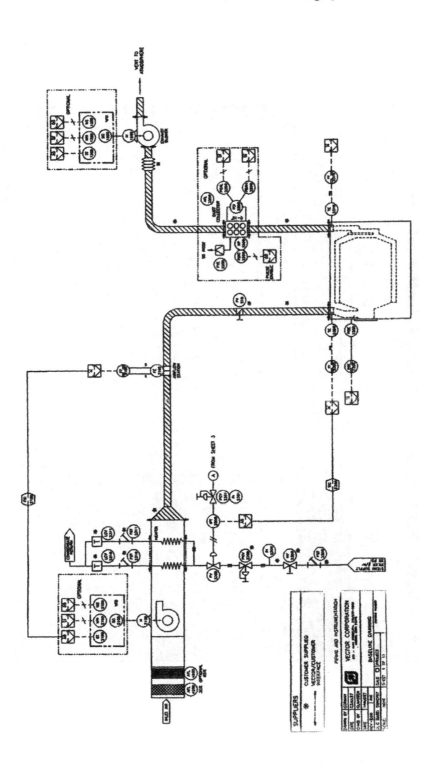

Appendix C: An Example of a Process Screen Utilizing the Grid Method

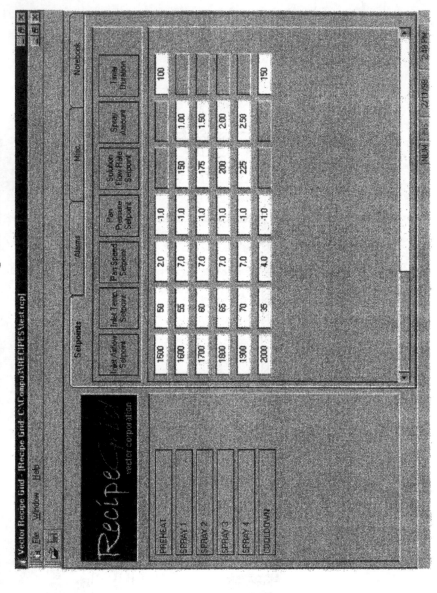

*Appendix D: An Example of a Computer Screen Showing an Object–
Oriented Recipe Management System*

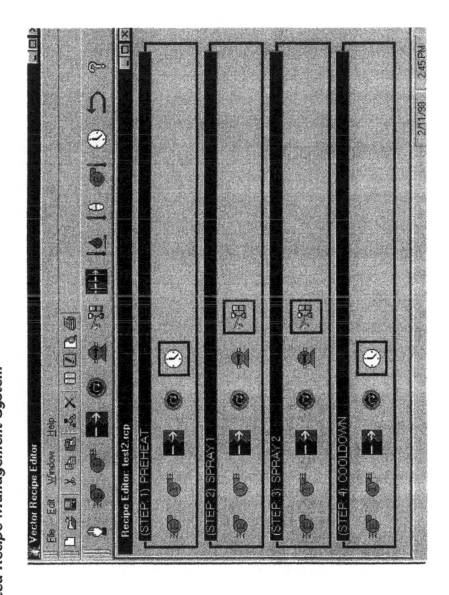

6

CLEANING AND VALIDATION OF CLEANING FOR COATED PHARMACEUTICAL PRODUCTS

William E. Hall

Hall and Associates

The equipment used in manufacturing coated pharmaceutical products must be cleaned. These cleaning procedures must be validated to demonstrate the adequacy and consistency of the cleaning process. The basic requirements for cleaning and cleaning validation are detailed in the current Good Manufacturing Practice (cGMP) regulations. The lack of documented, validated cleaning procedures has resulted in a significant number of inspection failures. One of the most important inspections of a pharmaceutical company occurs when a company desires to place a new product on the market. The U.S. Food and Drug Administration (FDA) inspection that precedes this event is known as a preapproval inspection (PAI). Table 6.1 contains the data that demonstrates the failure rate for pharmaceutical companies during the seven-year history of the PAI program (Phillips 1997). Not all of the failures reported in Table 6.1 were due to cleaning issues; however, a significant number of the failures were due to shortcomings in the validation of the cleaning processes.

Table 6.1. History of Preapproval Inspection Failures, 1991 to 1997

Year	Number of Inspections	Number of Withhold Approvals	Percentage of Approvals Withheld
1991	437	264	60%
1992	466	182	39%
1993	359	135	38%
1994	325	94	29%
1995	371	111	30%
1996	395	71	18%
1997	429	92	21%

Similar failures have occurred in FDA inspections of foreign pharmaceutical manufacturing facilities. Estimates of the percentage of failed foreign inspections range as high as 40 percent (Ellsworth 1996). Again, this is due to a variety of reasons, cleaning being one of the major causes.

RESPONSIBILITY FOR CLEANING

Until recent years, cleaning was often ignored and became an orphan responsibility with no clear ownership. Cleaning procedures were often developed very late in the product development schedule and usually by the production unit. However, now the responsibility for developing cleaning processes usually resides with the research and development (R&D) department because of their knowledge of the product formulation. As such, cleaning procedures are developed very early and, in today's regulatory climate, it is common practice for the cleaning procedure to be developed concurrently with the product formulation. Although stimulated by regulatory requirements, cleaning has now been placed on a much more scientific footing. A company no longer uses the same cleaning procedure for all products manufactured; a specific cleaning procedure is developed for each and every product.

COMPONENTS OF A CLEANING PROGRAM

A good cleaning program has several components:

- Appropriate, detailed cleaning procedures.
- A comprehensive personnel training program.
- Visual examination of equipment.
- Validation of cleaning procedures.
- A detailed documentation system.
- Change control.

Each of these aspects will be addressed in this chapter.

THE CLEANING PROCESS

The general objective of the cleaning process is to reduce cross-contamination from one product to another product or from one batch of a product to a subsequent batch of the same product. Cleaning procedures for pharmaceutical products may be manual, semiautomatic, or fully automatic. Although current trends are to automate all processes as fully as possible, most companies utilize manual cleaning in whole or in part to clean equipment or the surrounding production areas.

Fully automated cleaning processes are often referred to as clean-in-place (CIP) cleaning. CIP cleaning processes are usually very consistent, reproducible, and readily validated. Even though a CIP process may be used, it still requires validation. Samples must be taken, appropriately analyzed, limits determined, and all results documented clearly.

Equipment is usually not disassembled during a CIP process. This technology was originally developed by the dairy industry to clean the large tanks used to process milk and milk products. The cleaning solutions are delivered by a sprayball, which is essentially a spray device that may be a permanent part of the equipment or may be inserted only during the cleaning process.

The validation of CIP has become a science in itself, and there are numerous consultants who specialize in helping companies employ this technology. Cleaning cycles must be developed that define the concentration of cleaning agent, the temperature of the wash water, time of washing, time of rinses, and the number of rinses. Once the CIP cycles are developed and validated, the cleaning processes should deliver consistent results as long as no changes are made to any of the cleaning parameters.

Another type of cleaning is clean-out-of-place (COP). This type of cleaning process involves the disassembly of equipment, which is then taken to a central washing station. The washing equipment is often centrally located and may serve multiple manufacturing areas. While this arrangement may provide an efficient use of facilities, it also presents an opportunity for cross-contamination from other products. COP equipment must be validated to demonstrate that the proper temperature is reached, cycle times are met, and that detergent is dispensed in the proper amounts. An inherent advantage of this type of cleaning is that the equipment is disassembled during each cleaning event and thus can be visually inspected during the reassembly process. The major disadvantage to COP is that additional resources are required to develop and validate the cleaning cycles.

Almost all pharmaceutical facilities have at least some manual cleaning processes in their operation. Manual cleaning means that equipment is washed and scrubbed by hand using brushes, sponges, or cloths. While some may argue that manual cleaning processes are not validatable, the consensus opinion of experts is that manual cleaning processes can and must be validated. The key issue for manual cleaning is the level of detail in the descriptions of cleaning procedures (i.e., Standard Operating Procedures, SOPs) and the quality of the operator training program. Most difficulties with manual cleaning processes are due to inadequate training and variation in interpretation among workers as to what the written procedure states. One way of preventing misinterpretation is to have very detailed and specific cleaning procedures. Terms/phrases such as "approximately," "clean until no residue is apparent," "warm water," "hot water," and many other terms are known as "weasel words" and should be avoided in written cleaning procedures. Cleaning procedures should precisely indicate what is to be done, even to the point of specifying items such as the time glassware should be brushed or rinsed, the number of rinses, and the temperature of the water to be used in cleaning. Only when nothing is left to interpretation can the cleaning process be reproducible and consistent day after day.

In the past, some companies have written cleaning procedures that involve cleaning the equipment until it achieves some level of visual appearance or when a sample subjected to chemical analysis is below a specified level. The FDA refers to this type of testing as "test until clean," but this approach is not favored by the FDA. To a regulatory agency, this approach indicates that a company does not actually know what is required to clean the equipment—it has an undefined or ill-defined cleaning process. Even worse, it indicates that the company does not actually know the capability of the cleaning process; otherwise, the company would "cut to the chase" and simply clean for a specific time and then stop.

DEDICATED VERSUS MULTIPRODUCT EQUIPMENT

A major factor in the development of a cleaning program is determining whether the equipment is dedicated to the manufacture of a specific product or is also used for the manufacture of other products. In the case of dedicated equipment, the risk of cross-contamination of another product is essentially nonexistent. The cleaning procedures for dedicated product equipment must still be validated, but the extent of the validation will be much less and involve fewer test samples. Also, the protocols should not be as extensive, but the carryover of product to a subsequent batch of the same product may be very undesirable due to potential degradation of the material.

CLEANING AND THE TRAINING PROGRAM

The importance of training is well known within the pharmaceutical industry. Nowhere is this more apparent than in cleaning, especially for manual processes. Any company employing manual cleaning processes can expect to have its training program examined by FDA investigators.

It is important that the training program be current and well documented. Many companies do excellent "on-the-job" training but do not adequately document the training process. In the eyes of the inspector, if the training is not documented at the time it is done, it was simply *not* done.

It is also important that training be current. For some procedures that are performed infrequently, it would be easy to skip a step or not perform the step in a proper fashion because of unfamiliarity. Many companies perform retraining on an annual basis and update the training and the training records accordingly. Any time a procedure changes, it is necessary to retrain operators and to document the new training.

One indication that training may not be adequate is obtaining widely different results during a cleaning validation. If testing following cleaning by operator A indicated that high levels of residue remained on the equipment, whereas low levels of residue were observed after operator B, this probably indicates a training problem.

As a related issue, written copies of the cleaning procedure should be available in the production areas where cleaning occurs. In the past, cleaning procedures were often located in a supervisor's office or in a filing cabinet so they would not get wet during the cleaning process. This arrangement invites error. Written copies of the cleaning procedures can be laminated in plastic and placed on the room wall or attached to the equipment. In some cases, the cleaning procedure becomes an integral part of the batch records; thus consistency and control are achieved. Having the cleaning procedure as a part of the batch records ensures that the procedure will be in the manufacturing area and guarantees that a second person will witness the cleaning and document the cleaning with a second signature.

VALIDATION OF CLEANING PROCEDURES

It is not sufficient to have only a written cleaning procedure. The cleaning procedure must also be validated. Validation is a term that is often misunderstood. Basically, validation consists of preparing a written experimental plan that will demonstrate that a procedure actually does what it is intended to do and that it does so consistently every time. For the validation of a cleaning procedure, it is necessary to devise an experimental plan that will result in documentation demonstrating that the cleaning procedure reduces product and microbial residues to an acceptable level. During the validation process, the cleaning procedure is performed three times, and the clean equipment is tested each time for the presence of any remaining residues. All three tests must be successfully passed to demonstrate the consistency of the cleaning process.

The initial task in developing a cleaning validation program is to determine what products and processes will be evaluated. Ideally, one should validate the cleaning of all equipment surfaces that are likely to contact product. For a company manufacturing hundreds of products, the combinations and permutations of products and equipment would require literally thousands of cleaning validation studies and tens of thousands of analytical samples. For such a situation, a valid approach would be to divide the products into potent and nonpotent categories. This categorization is somewhat subjective since all products must have some degree of potency to be effective. However, there are products such as penicillins, cephalo-sporins, cytotoxic agents, and allergenic substances that are much more potent and more serious if present as a contaminant. These products are usually manufactured in dedicated equipment, where any cross-contamination would not have consequences as serious as would be the case if they were manufactured in multiproduct equipment. Another matter to consider in the validation of cleaning procedures for an entire product line and facility is the grouping of products and selection of a worst-case product for cleaning valida-tion purposes.

In order to understand the worst-case approach, consider the following example. Suppose that four coated tablet products are manufactured in the same manufacturing equipment. Assume that the following information about the four products in the group is known:

- Product 1 is a sugar-coated aspirin tablet.

- Product 2 is a sugar-coated pseudoephedrine tablet.

- Product 3 is a film-coated digoxin tablet.

- Product 4 is a sugar-coated triprolidine tablet.

Of the four products, the product having the most potent active in-gredient is the digoxin tablet. The digoxin would also be the least soluble (i.e., the most difficult to clean) of the active ingredients, and, therefore, the worst case.

Another example might be represented by a grouping in which all of the products had the same active ingredient but the coating process involved different materials. For example, consider a prod-uct grouping composed of the following three hypothetical prod-ucts:

- Product 1: 15 mg pseudoephedrine sugar-coated tablet.

- Product 2: 30 mg pseudoephedrine sugar-coated tablet.

- Product 3: 120 mg pseudoephedrine wax-coated extended release tablet.

In this example, product 3 would be the worst-case product for cleaning purposes because (1) the wax coating would be more diffi-cult to dissolve and, therefore, the hardest to remove and (2) this product has a greater quantity of active ingredient and would, therefore, present the greatest cleaning challenge and the greatest risk of toxic (potent) residues.

In the above examples, it would be easy to prepare a logical rationale for the grouping and choice of the worst-case representative of the group. A protocol would then be prepared and implemented for the worst-case product, with the stated purpose that its validation will satisfy the need for the validation of all products in the group and that no additional cleaning validation will be performed on the other products in the group. In some cases, there may be no single worst-case product and two or more worst-case products may be selected. The important point is that there is a scientific rationale to defend the choice(s); the approach must be logical, practical, and achievable. The nature of the coating process itself could be used as the basis for product groupings. For example, it might be feasible for a given company to group products as sugar-coated products, film-coated products, and gelatin-coated products. This approach has been utilized by contract manufacturers who specialize in coated products (e.g., 1,000 to 4,000 different products). Almost infinite testing would be required to carry out product-specific cleaning validation studies for this many products; the grouping concept allows the company to reduce the workload to an achievable level.

It is important to not extend the grouping too far such as having all coated products covered by a single worst case, unless all coated products are similar in composition and the equipment used to apply the coatings is the same.

SOURCES OF CONTAMINATION

There are three main types of contamination:

- Chemical contamination

- Bacterial contamination

- Process materials contamination

Chemical contamination can result from the active ingredient, excipients, degradation products, or cleaning agents. Most of the present focus for coated products is on the active ingredients and the cleaning agents. These two groups must have a formal cleaning validation package, consisting of at least three sets of cleaning data.

The presence of objectionable microorganisms is not allowed by regulation. Since many coating solutions are aqueous in nature, this is a potential source of viable bacterial contamination. Any water that is present in dead legs or other difficult-to-access equipment parts should be eliminated either by equipment design, equipment modification, or cleaning procedures that will ensure that equipment is rendered dry by blowing air or nitrogen over/through the equipment prior to storage. This is best dealt with by procedures that prevent bacteria from entering the product. The quality of raw materials, including rinse waters, should be controlled. Natural materials such as gelatin, which are commonly used in preparing coating solutions, are a potential source of bacterial contaminants. All avenues for prevention of microbial growth, such as using only freshly prepared solutions, should be employed.

Other materials that also may have product contact and have the potential to cause contamination of subsequent products include lubricants, gaskets, porous tubing, glass containers, filters, stainless steel equipment, carbon filters, aluminum equipment, and filtering aids. One of the basic premises of a cleaning program is that confirmatory visual examination be performed after every cleaning, not just during cleaning validation studies. One of the main reasons for this requirement is that tests are not done for every possible contaminant. For example, particles shed from stainless steel equipment may not be detected. However, if visual examination is performed, these particles would probably be detected. Visual examination is also useful for evaluating equipment surfaces that are difficult to clean.

SAMPLING

There are many different types of sampling techniques for evaluating cleaning: swab sampling, solvent rinse sampling, rinse water sampling, placebo sampling, sampling of the next product manufactured in the equipment, direct surface monitoring, coupon sampling, and combinations of these types. Even within a category for a given sampling method, there are subtypes of sampling. For example, there are dry swabs and wet swabs; the latter are moistened with water or another solvent. Those sampling methods that are acceptable to the FDA are swab sampling, rinse sampling, and solvent rinse sampling (since these terms are referred to in the various inspection guides utilized by FDA inspectors).

The FDA has a strong preference for swab sampling, since it is believed that some residues require mechanical or physical action in order to remove the residue and that rinse samples might give a false indication that the equipment was clean. When product contact surfaces lend themselves to easy access by the sampler, swab sampling is easily accomplished. However, there are product contact surfaces that do not lend themselves to easy access, such as the inner surfaces of hoses, transfer pipes, and small intricate equipment such as spray heads, micronizers, microfluidizers, and numerous other examples. Tanks may not have ports large enough for entry for sampling purposes. For wire screens, sieves, or brushes, swabbing simply is not appropriate because of the nature of the material. It is then reasonable and acceptable to use either solvent or water rinse sampling. Because residue tends to be trapped in the spray equipment used for coated products, a better sampling program would involve soaking the disassembled spray nozzle in the solvent for which the residue is known to be soluble.

The main concern of rinse sampling is that if large volumes of rinses are used, the residue can be diluted to the extent that it will not be detected by the analytical method. To overcome this potential difficulty, sampling volumes should be kept as low as possible or the rinse samples may be concentrated by evaporation of some or all of the solvent after sampling.

Regardless of the sampling methodology, a sampling protocol should be developed that identifies the size and location of all samples. Normally, a schematic diagram of the equipment is prepared, and the sampling locations may be marked directly on the diagram. The spray nozzles of spray-coating equipment should receive special sampling since they are at a terminal location in the manufacturing process. Any contamination in a spray nozzle could conceivably be delivered to the first few units of the next product manufactured. Thus, the spray nozzle is a critical sampling site. Other locations may be less critical, since any contaminant would be uniformly distributed in the next product by nature of the mixing in the manufacturing process.

The sampling locations should include all locations that are known to be difficult to clean. The sampling protocol should indicate what type of samples (rinse, swabs, etc.) will be used and should include the details of exactly how the samples will be obtained. Details that should be addressed in the sampling protocol include the volume of rinses, the area sampled, the number of strokes of the swabs, the direction of swabbing (i.e., horizontal, vertical, or both), and how the samples should be labeled and transported to the analytical laboratory.

For CIP systems, it is advisable to disassemble the equipment during the cleaning validation for sampling purposes, even though the equipment is not normally disassembled during use. The disassembly not only facilitates sampling, but also allows the sampler to examine the inner product contact surfaces visually to determine if there is gross contamination that remains undetected by the sampling and analytical technique.

The importance of the sampling process cannot be overemphasized. Even the most sophisticated instrumentation and skilled analyst cannot compensate for improper or inadequate sampling.

ANALYTICAL METHODS

Analytical methods used for cleaning samples must be carefully chosen for the specific situation required. The most common error is in choosing a method that is not sensitive enough. The sensitivity must be low enough to detect residue at the limit set for the residue. An error associated with the sensitivity is the assumption that "none detected" is equal to zero. It is important to remember that "none detected" is not equal to zero, but instead is equal to the sensitivity of the analytical method as reflected in either the limit of detection (LOD) or the limit of quantitation (LOQ). Another factor to remember is that the analytical method used for testing cleaning samples must be validated itself—a concept often referred to as methods validation.

Because of their codependency, it is often highly desirable to develop analytical and sampling methodologies concurrently. If the sampling will be done by swabbing, then one of the first activities for the analytical department should be to determine the percent recovery from the surface of the equipment. This is usually done by spiking known amounts of the expected residue on surfaces of the same composition (usually stainless steel) as the equipment to be sampled. The recovery would be simply defined as

$$\text{percent recovery} = \frac{\text{amount detected}}{\text{amount spiked onto surface}} \times 100$$

The question often arises as to what is an acceptable percent recovery. There is no regulatory requirement for recovery; indeed, the range of values reported varies greatly. Values as low as 15 to 20 percent have been reported by biotechnology companies. This is neither good nor bad; it is a function of the nature of the materials, the levels of residue encountered, and may be the maximum attainable for residues such as poorly-soluble materials (e.g., proteins). For very soluble materials, the percent recovery may be as high as 99.9 percent. Most typical recoveries fall somewhere between these extremes, and typically are in the range of 50 to 70 percent.

One way to enhance poor recoveries is to use a piece of filter paper saturated with solvent as the swab. This wetted filter paper may be placed directly on the equipment surface, rubbed on the back with a glass rod or a rubber "policeman," and allowed to stay in contact with the equipment for several minutes. This method combines the physical action of swabbing with the solvation action of the solvent and may improve recoveries in some situations. A list of analytical methods typically used for cleaning validation samples is given in Table 6.2 and also indicates whether the method is specific or nonspecific in nature. Specificity or nonspecificity is significant since several regulatory guidelines refer to the use of specific analytical methods. While many of the simpler methods, such as visual, pH, conductivity, and TOC are nonspecific in nature, they still

Table 6.2. Analytical Methods for Cleaning Validation

Analytical Method	Specific	Nonspecific
Visual examination*		✔
Gravimetric analysis		✔
pH		✔
Conductivity		✔
Microscopy		✔
Titration		✔
Thin layer chromatography (TLC)	✔	
Lowry protein	✔	
SDS–PAGE	✔	
Capillary zone electrophoresis	✔	
ELISA	✔	
Ion chromatography	✔	
Fourier-transform infrared (FTIR)	✔	
Near infrared (NIR)	✔	
High performance liquid chromatography (HPLC)*	✔	
Total organic carbon (TOC)*		✔

*Indicates the most commonly used analytical methods for cleaning validation

render valuable information relative to the level of cleaning and the presence of any possible contaminant. Some of the nonspecific methods are extremely sensitive and rapid. Due to these properties, they can often be very effective in evaluating cleaning as well as offer a valuable monitoring application since they lend themselves to in-line or on-line application.

For several years, heat distillation stills were equipped with conductivity monitors that automatically shut down the still or drained water when the water did not meet the standard programmed into memory. An extension of this technology to include on-line monitoring of pH and TOC in addition to conductance is presently used in several new facilities. Using these nonspecific analytical methods in tandem has considerable merit when applied to cleaning samples. Even though each method has individual disadvantages or limitations, together they cover a vast array of potential contaminants. The conductance measurement would detect any inorganic, ionic contaminant, the pH measurement would detect any residue having acidic or basic character, and the TOC measurement would detect any organic contaminant.

Although originally there was great reluctance by FDA investigators to accept any analytical method that was not product specific in nature, there has been some moderation in this position. In some cases, nonspecific methods may actually give a much better indication of the clean condition than a product specific method. An example of this is in biotechnology manufacturing, which may involve more than 100 components (buffers, salts, media, products). Although it may be impractical to test for all 100 components by individual, product-specific methods, the status of the equipment can be quickly evaluated by TOC. If only a few product-specific assays were performed, it is quite possible to have anomalous results that seem to indicate that the equipment is clean, when, in fact, the equipment was not clean because it was not assayed for the specific contaminant(s) present. Another stimulation to nonspecific analytical methods was given when the U.S. Pharmacopeia (USP) indicated that it plans to adopt TOC as an alternative for testing water for oxidizable substances.

Visual Examination

Although a nonspecific method, it is probably the most popular and easy to use analytical method of all. Some companies have carried out studies to make this technique quantitative for their particular application. In contrast to other analytical methods, this method gives the observer the most complete and immediate indication of the condition of cleanliness in the equipment. For situations where the entire product contact surface can be observed (e.g., a large manufacturing tank), the entire surface can be evaluated visually. This simple method bypasses the difficulties of taking a finite sample from a limited location or series of locations. It also does not suffer from recovery difficulties as does swab sampling or rinse sampling. Another benefit of visual examination is that it will allow the observer/operator to detect gross amounts of contamination concentrated in a small area, which could go undetected with "normal" sampling programs.

Various modifications of visual examination have been utilized in the pharmaceutical industry. Visual examination has been enhanced by the use of ultraviolet (UV) light when a potential contaminant has fluorescent properties. For areas that are difficult to access, such as behind tank baffles or inside transfer pipes, companies have used various creative approaches, including fiber optic probes and video cameras.

The visual examination of equipment should be a component of all cleaning validation programs regardless of whether additional testing methods are used. It is actually quite common practice to perform both visual examination and chemical/biological testing of cleaned equipment during the validation process.

Analytical Techniques for Biotechnology Cleaning Validation

There is a family of analytical techniques that are widely used in cleaning validation for biotechnology products, including Lowry protein, sodium dodecyl sulfate–polyacrylamide gel electrophoresis (SDS–PAGE), capillary zone electrophoresis, enzyme-linked immunosorbent assay (ELISA), and TOC analysis. These methods are particularly useful and appropriate since most of the products and potential contaminants are protein in nature.

High Performance Liquid Chromatography

High performance liquid chromatography (HPLC) has been the mainstay of pharmaceutical analysis for many years. Most recently, analytical researchers have developed a new generation of detectors for these instruments, which has extended the range of applicability to an even wider array of materials. The new detectors, known as evaporative light-scattering detectors (ELSDs), are more universal in their response (i.e., they do not require a chromophore group on the molecule as do UV detectors). For ELSDs, all compounds produce similar responses, and there are no baseline drifts due to mobile phase effects. These developments in analytical technology should make this technique even more useful in the years to come.

Microbial and Endotoxin Detection

Microbial and endotoxin contamination has become and will continue to be even more significant both in dosage form manufacturing and bulk pharmaceutical chemical (BPC) production. Even though these testing methods were not listed in Table 6.2, they are important. Such methods are especially important for detecting bacterial contamination of products derived from natural sources, biotechnology products, and aqueous-based processes.

Total Organic Carbon Analysis

The newest member on the analytical scene is TOC analysis. By this method, the carbon atoms in the analyte are oxidized to carbon dioxide, which exists in aqueous solution as the bicarbonate anion:

$$\text{carbon-containing residue} + \text{oxidizing agent} \rightarrow HCO_3^-$$

The bicarbonate anion is then detected by various types of detectors, typically infrared or conductimetric detectors. The essential analytical instrumentation is basically very simple but has been automated with microprocessors that determine when the sample is completely oxidized, subtract the inorganic carbon present (carbon in the water itself as carbonate or dissolved carbon dioxide), and automate the analytical process so that multiple samples can be run unattended.

TOC offers great promise for verifying and validating the cleaning process. It is extremely sensitive (one instrument supplier claims 50 ppt, i.e., 50 parts of carbon per 1,000,000,000,000 parts of solution). It also involves minimal development time compared to other analytical methods, and the instrument run time is rapid, thus enabling a laboratory to generate literally hundreds of data points from an unattended instrument running overnight.

There are two disadvantages to TOC analysis. The first has already been referred to—this method is a nonspecific analytical method. The second major disadvantage is that the material to be analyzed must first be dissolved in water, which requires the substance to have at least a minimal aqueous solubility. It remains to be determined whether these disadvantages will prevent this analytical method from being utilized in an almost dominant fashion for cleaning sample analysis. In any event, the technique is already being used by many companies (Baffi et al. 1991; Jenkins et al. 1996).

Two possible applications of TOC to testing for cleaning are readily apparent. (1) The technique is useful as a screening tool, whereby the equipment is sampled, samples are assayed by TOC, and the resulting data are used to screen the equipment to determine the general level of cleaning and identify the most difficult-to-clean locations. The initial screening by TOC could then be followed by product-specific assays, with particular attention to the areas identified as hard to clean. (2) The TOC data may be useful for cleaning validation purposes. In order to use the data for cleaning validation, it is necessary to prepare a scientific rationale that essentially states that all carbon residue detected will be assumed to have resulted from the most potent material present, normally the active ingredient. If the actual cleaning samples give results that are less than the limits established for the most toxic material, then the actual identity of the contaminant is not necessary, since the worst-case rationale was used. If the actual results exceed the previously established limits, then a product-specific assay could be used to establish whether the residue was due to active ingredient(s), excipients, cleaning agents, preservatives, or other ingredients in the formulation. The only case where this argument would be scientifically flawed is if the potent material (i.e., active) was not an organic (i.e., carbon-containing) compound or if the active could be entrapped by other insoluble ingredients and thus not available to be dissolved and detected analytically. It should be noted, however, that this is a potential problem with all analytical methods, specific as well as nonspecific.

LIMITS AND ACCEPTANCE CRITERIA

Regardless of the cleaning approach or strategy, the question inevitably arises as to what is clean enough or "how clean is clean"? There is no single, comprehensive answer to this question. It is quite clear that a detection level of zero residue is not practical due to the tremendous sensitivity of analytical technology.

The cleaning limit, although very important, is only a single component of the entire cleaning validation program or protocol. Other parts of the program are the requirements that equipment should be visibly clean, that training records of cleaning personnel should be accurate and current, and that the cleaning procedures should reflect the procedures actually used by the operators to clean the equipment.

The most important aspect of arriving at a cleaning limit is the journey itself, not the destination. Several companies are still using a limit of "x" ppm because they have heard a number mentioned at a meeting and have decided to apply it to their operation. Needless to say, this is not and will not be acceptable to the FDA. The journey referred to is the scientific rationale that provides the basis for the numerical limit itself.

A good scientific rationale should be logical, practical, verifiable, and achievable. The numerical limit should be based on one or more of the following:

- Therapeutic dosage level of the active(s).

- Toxicity of the material.

- Solubility of the potential residue.

- Difficulty of cleaning.

- How the products will be used.

- The nature of other products made in the same equipment.

- The batch size of other products made in the same equipment.

For a finished pharmaceutical dosage form or a drug substance, the limit is often based on allowing not more than a fraction of a therapeutic dose to be present in subsequent products. Often, for oral dosage forms (tablets, capsules, caplets, etc.), a fraction of the smallest therapeutic dose, for example, 1/1,000th, is used as the numerical limit. The 1/1,000th in this case becomes a "safety factor." The safety factor should be a measure of a reasonable degree of risk for the given situation. The reasonable degree of risk will be different for BPCs and dosage form manufacturing. The risk due to contamination is different for different dosage forms. For example, a given amount of contamination would present a more serious risk in an intravenous dosage form than in an oral dosage form such as a tablet, caplet, capsule, or syrup. In a similar fashion, contamination of an oral dosage form tends to be more serious than contamination of a topical dosage form.

To carry the concept of risk into the limits determination for cleaning residues, there is no reason that risk cannot be related to the nature of how the product will be used. For calculations of risk for dosage form manufacturing based on therapeutic doses, it may be feasible to adjust the safety factor based on the dosage form and to develop a continuum of safety factors (Table 6.3).

It is apparent that risk, as expressed in the safety factor, is justifiably different for different dosage forms. The risk is higher for research compounds because little may be known about the toxicity of the material or its effect in a diseased body. Therefore, a greater safety factor is applied (i.e., a smaller fraction is allowed to carryover), and the limit will be lower for these cases. At the other end of the continuum is the dosage form of lowest risk (i.e., the topical dosage form). Since there is far less risk for contamination to cause medical problems for this dosage form, the safety factor is appropriately less (i.e., a larger number). Of course, the majority of the dosage forms will fall somewhere in between these two extremes (Table 6.3).

Calculation of Limit Based on Smallest Therapeutic Dose

As an example of a simple limit calculation, assume product A is manufactured and the equipment is subsequently cleaned before other products are manufactured. Assuming that the product will ultimately be used as an oral tablet and that the smallest therapeutic dose is 100 mg, a safety factor of 1/1,000 is applied. This means that the next product should contain not more than:

$$100 \text{ mg} \times 1/1,000 = 0.1 \text{ mg per daily dosage}$$

If it is known, for example, that product B will have a maximum daily dose of 1,000 mg (e.g., 10 tablets each containing 100 mg of

Table 6.3. Safety Factor Continuum

Dosage Form	Safety Factor
Research compound	1/100,000 to 1/10,000
Intravenous products	1/10,000 to 1/5,000
Ophthalmic products	1/5,000
Oral dosage forms (tabs, caps, caplets)	1/1,000
Topical products	1/100 to 1/10

active) and a batch size of 300 kg of active ingredient, then it is possible to calculate the limit using a simple proportion as follows:

$$300 \text{ kg} = 300,000,000 \text{ mg}$$

$$\frac{0.1 \text{ mg}}{1,000 \text{ mg}} = \frac{x \text{ mg}}{300,000,000 \text{ mg}}$$

$$x = 30,000 \text{ mg}$$

It is important to note that the 30,000 mg limit appears to be quite large; however, this is the total residue allowed for all manufacturing and packaging equipment. This is only one simple example of calculating a limit based on smallest therapeutic dose.

Some companies use a worst-case approach for this calculation. In the above example, the calculation would be modified by using the smallest batch size of any product made in the same equipment and the largest daily dose of any product made in the same equipment. This allows a single limit to be set, instead of having different limits depending on the parameters of the next product. For the above example, if the smallest batch size for any other product made in the equipment is 100 kg of active ingredient and the largest daily dosage of any other product made in the same equipment is 1,500 mg of active ingredient, then the limit calculation would be as follows:

$$100 \text{ kg} = 100,000,000 \text{ mg}$$

$$\frac{0.1 \text{ mg}}{1,500 \text{ mg}} = \frac{x \text{ mg}}{100,000,000 \text{ mg}}$$

$$x = 6,667 \text{ mg}$$

It is important not to interpret these calculations such that one method is correct and the other incorrect. By the first method, a different calculation would be required for each and every product that followed product A. Thus, there would be a different limit if product B followed product A, if product C followed A, if product D followed product A, if product A followed B and so forth, for every possible combination and sequence of manufacturing events. This would become very unwieldy to manage and, thus, many companies choose the second approach (i.e., of using smallest batch size and largest daily dose for all products made in the same equipment).

One obvious limit to this method of calculation is that in order to use it, there must be an established therapeutic dose. Not all potential contaminants have therapeutic doses. For example, there are no therapeutic doses for precursors, by-products of chemical synthesis, and cleaning agents (detergents). Therefore, a method of calculating limits is needed that is based on some parameter other than therapeutic dose. One method that can be used in these instances is based on the toxicity of the material.

Calculation of Limit Based on Toxicity

This method of calculation is based on using animal toxicity data to determine limits and is particularly suited for determining limits for materials that are not used therapeutically. This method is based on the concepts of acceptable daily intake (ADI) and no observed effect level (NOEL) developed by various scientists in the Environmental Protection Agency (EPA) (Dourson and Stara 1983), the U.S. Army Medical Bioengineering Research and Development Laboratory (Layton et al. 1987), and the Toxicology Department at Abbott Laboratories (Conine et al. 1992).

Basically, these workers were attempting to determine the amounts of chemicals that the human body could ingest on a daily basis without undue risk and toxicity. In the process, they found that the ADI could be calculated from the toxicity of the materials expressed as an LD_{50}. Such data are widely available on Material Safety Data Sheets and other references where toxicity data can be found.

NOEL is calculated from the LD_{50} by the following mathematical relationship:

$$NOEL = LD_{50} \times 0.0005$$

where 0.0005 is a constant derived from a large toxicology database. Once NOEL is known, then the ADI can be calculated by the relationship

$$ADI = NOEL/SF$$

where SF is an appropriate safety factor. Finally, the maximum allowable carryover (MACO) can be calculated from the following relationship:

$$MACO = ADI \times B/R$$

where B is defined as the smallest batch size of any other product made in the same equipment and R is the largest normal daily dosage of any product made in the same equipment.

For example, consider a fictitious chemical substance, chemical X. If it is assumed that the toxicity, batch size, and dosage information is known, then MACO can be calculated as follows:

LD_{50} = 419 mg/kg (oral) and 85 mg/kg (IV)

Smallest batch size made in same equipment (B) = 40 kg

Largest normal daily dosage (R) = 300 mg

NOEL = 419 mg/kg $\times$ 0.0005 = 0.2095 mg/kg/day

for a normal adult of 70 kg,

NOEL = 0.2095 mg/kg $\times$ 70 kg = 14.665 mg

Using a safety factor of 100 (for the oral route),

ADI = 14.665/100 = 0.147 mg

$$MACO = \frac{0.147 \times 40{,}000{,}000 \text{ mg}}{300 \text{ mg}} = 19{,}600 \text{ mg or } 19.6 \text{ g}$$

Similar calculations for the IV route of administration are as follows:

$$NOEL = LD_{50} \times 0.0005 =$$
$$85 \text{ mg/kg} \times 0.0005 = 0.0425 \text{ mg/kg/day}$$

For the typical 70 kg adult,

NOEL = 0.0425 mg/kg/day $\times$ 70 kg = 2.975 mg/day

ADI = 2.975/5000 = 0.000595 mg/day

(Note: The 5,000 represents a safety factor for the IV route of administration.)

$$MACO = \frac{0.000595 \times 40{,}000{,}000 \text{ mg}}{200 \text{ mg}} = 119 \text{ mg}$$

This calculation illustrates a couple of additional points. First, the MACO calculation will utilize different LD_{50} values depending on the route of administration of the other products manufactured in the same equipment. If all of the products manufactured in the equipment were used orally, then the limit used would be 19.6 g. However, if any of the products made in the same equipment were to be eventually incorporated into an IV dosage form, then the limit would be 119 mg (i.e., the most conservative of the two calculations.

Another important aspect of limits calculations is that the values calculated represent the total amount of allowable residue on *all* pieces of equipment in the manufacturing "train." Often, for practical and logistics purposes, it is necessary to divide, or prorate, the limit among the various pieces of equipment.

Table 6.4 illustrates how the limit is prorated for a specific manufacturing setup. It is apparent that the total limit is divided or proportioned based on its percentage of the total surface area.

If rinse sampling is used and the entire equipment is rinsed, then the limit can be used for the individual equipment item. However, if the equipment will be sampled with a swab, it is necessary to factor the limit even further. For example, if 6 areas of the manufacturing tank will be sampled by swab sampling and each swab represents an area swabbed of 12 in. by 12 in., then the total area

Table 6.4. Dividing a Total Residue Limit Among Various Equipment Items

Equipment	Surface Area (ft²)	% of Total	Limit (Oral) (grams)	Limit (IV) (grams)
Manufacturing Tank	23	6.34	1.24007	0.0075
Transfer Tank	23	6.34	1.24007	0.0075
Holding Tank	98	27.03	5.28378	0.0322
Centrifuge	45	12.41	2.42623	0.0148
Pre-Dryer	116	31.99	6.25428	0.0381
Dryer	28	7.72	1.50965	0.0092
Prefilter	27	7.45	1.45574	0.0089
Line Filters	2.6	0.72	0.14018	0.0009
Totals	362.6	100	19.5500	0.1190

swabbed is 6 ft^2. (Note: The total area of the equipment was 23 ft^2.) The total allowable residue for all 6 swabs would be as follows:

$$\text{limit for total area swabbed (oral)} = 6/23 \times 1.24007 = 0.3235 \text{ g}$$

$$\text{limit for total area swabbed (IV)} = 6/23 \times 0.0075 = 0.002 \text{ g or 2 mg}$$

To determine the residue allowed per swab, it is necessary to divide these results by 6:

limit for single swab (oral) = 0.3235/6 = 0.0539 g

limit for single swab (IV) = 0.002/6 = 0.0003 g or 0.3 mg or 300 mcg

CLEANING VALIDATION DOCUMENTATION

Validation of the cleaning process is the accumulation of testing documentation that demonstrates that the cleaning process consistently reduces the levels of residues remaining on equipment to previously determined acceptable levels. It is important to recognize that validation is synonymous with documentation. Once a cleaning method has been developed for a specific coated product, the process of validating the cleaning process may begin. The documentation is extremely important and must be retained by the company as long as the cleaning process is used. Unlike batch record documentation, which may be discarded when product batches reach their expiration date, the cleaning validation documentation can be discarded only when the cleaning process is changed and revalidated. The cleaning process may be provided for in an overall company Master Plan or may be a stand-alone document. In either case, a protocol should be written that describes the experimental plan or blueprint that will be used to carry out the sampling and testing and provide for the accumulation of data that proves that the cleaning procedure is effective in a consistent fashion.

It is also important to recognize that, for validation purposes, the cleaning of equipment involved in the coating process is only one component of the cleaning validation for a coated product. The coating process represents only a single step in the complete manufacturing procedure. The overall manufacturing procedure can be represented as follows:

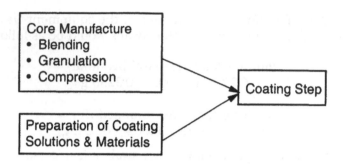

This is significant because cleaning limits are normally based on the total amount of residues for all manufacturing steps. This so-called "train concept" is necessary since each manufacturing step can contribute residue to the next product, and the total amount of potential carryover contamination is the numerical sum of all residues for all manufacturing steps, not just the coating step. The total amount of residue could be represented as

Residue on blending equipment +
Residue on granulation equipment +
Residue on compression equipment +
Residue on coating equipment +
Residue on packaging equipment

Total Residue = sum of all residues =
total potential carryover to next product

The documentation resulting from cleaning validation may be voluminous, but it can be divided into two categories: Protocols and final validation reports (often referred to as packages)

Protocols

The protocol is the experimental plan that will be followed to demonstrate the capability of the cleaning process. This experimental plan usually contains some or all of the following elements:

Scope

The scope is a written description of the products and processes to be covered by the experiments outlined in the protocol.

Objective

The objective is a statement of what is to be accomplished by the experiments. It is typically a short general statement that expresses the purpose of the study (e.g., to demonstrate that the cleaning procedure will successfully and consistently reduce the levels of residue to a predetermined level of acceptability).

Description of Process

A brief description of the manufacturing process is often included in this section, as well as a detailed description of the cleaning process or a reference to the number of the SOP or other written procedures. Any equipment used in the manufacturing process and any special auxiliary cleaning equipment, such as high pressure rinsing equipment, cleaning agents, and washing machines, should be given here. It is also appropriate to distinguish whether the cleaning process is CIP, COP, or manual.

Identification of Critical Parameters

For every process, including cleaning processes, there are parameters that must be controlled for the process to deliver consistent results. These parameters are known as critical parameters and must be both controlled and measured during the validation procedure. Examples of critical parameters for cleaning processes are the temperature of the cleaning solutions, the concentration of cleaning agent(s), the volume of cleaning solution, the extent of disassembly of the equipment, flow rates of cleaning solution (especially important for CIP systems), drying conditions to remove water from equipment, training of personnel, and storage conditions of cleaned equipment. This list is not complete, but it gives the reader a general idea of the types of factors that may differentiate a good cleaning process from a totally unacceptable cleaning process.

Test Functions

After the critical parameters are determined, the next step is to develop the actual tests that will be used to validate the cleaning process and determine if the cleaning procedure is adequate and validatable. An example of a test function is the visual examination of the equipment to verify that the equipment is visually clean. If equipment is not visually clean, then it cannot be considered clean, and there is little reason to proceed with chemical testing. Another value of physical examination is that often it is possible to detect residue visually when chemical analysis alone might fail because not all surfaces were swabbed or rinsed. Visual examination allows the quick evaluation of large and intricate surfaces and is the simplest of all tests.

Another important test function is training. It must be demonstrated during the validation process that personnel have been properly trained in the cleaning procedure. This is particularly important for manual cleaning processes, since there is the potential for significant differences in the way various individuals might interpret and actually carry out a given procedure.

There are also analytical test functions. These would be required for the quantitative determination of the amount of product and cleaning agent residues. As mentioned previously, it is very important that the analytical procedures for trace amounts of actives and cleaning agents be sensitive enough to detect the residues and that the analytical procedures themselves be validated.

Description of Sampling Process

A detailed description of what type of sampling (swab, rinse, or other) is to be used, how the sample will be obtained, and how the samples are to be stored until analysis is performed, is given in this section of the protocol. A sampling diagram is often included and indicates the location of all sampling sites, emphasizing the difficult to clean areas (hot spots and critical sites).

Description of Analytical Methods

A detailed description of the analytical process or a reference to an associated document that gives the details of the analytical method(s) is provided here. This section should also indicate that the analytical method itself is validated and, therefore, capable of detecting residue levels at the concentration established by the limits calculations. It should also address recovery studies that establish the effectiveness of the sampling process in removing residues and delivering them to the analytical instrument.

Limits and Acceptance Criteria

The limits and acceptance criteria are perhaps the most important and the most scrutinized portion of the entire protocol. The scientific rationale that supports the actual numerical limit as well as the other acceptance criteria is described here. It is important that the limit not be arbitrary but, instead, that it be based on medical dosage levels, medical effect levels, or the toxicity levels for the particular residue substance.

The rationale should also define what the basic approach is (i.e., whether a total residue [the so-called train concept] approach will be the basis for pass or fail or whether there will be separate, individual limits for each piece of equipment).

There may be multiple requirements for acceptability. For example, it may be a requirement that the total residue not exceed a certain level as well as that the concentration of contaminant not exceed a prescribed concentration (usually expressed in ppm).

There are usually additional requirements, such as that the equipment be visibly clean when examined with a "black" (UV) light. Examination under a UV light is useful for actives and excipients that are fluorescent. This makes the visual detection of residues remaining on the equipment easier.

A good rationale will also address assumptions or conditions that exist and may affect the risk of contamination. For example, if the equipment is dedicated completely or in part to the production of a single product or intermediate, the risk is much less than for a multiproduct situation, and this should be stated in the rationale. Many companies assume that if a residue were to occur in a particular piece of equipment, such as a mixer, it would be uniformly distributed in the following product. Assumptions of this nature should be stated in the rationale.

If there are calculations or equations used in determining quantitative limits, they should be described in detail, and each term and step in the calculation should be fully explained. In many cases, it is appropriate to include a sample calculation.

Documentation

The specific documentation that must be part of the final validation report is included here (e.g., original analytical records, recorder charts, reports, signed statements by the manager of the analytical department that the analytical methods are validated, and signed statements by production supervisors that personnel were properly trained in the use of the cleaning procedures.

Analyses and Conclusions

The data will be summarized in this section of the protocol and any deviations or failures will be fully explained or addressed. There should be at least three sets of data, since all validation requires at least three experiments (runs). In order to be acceptable, there must be three consecutive, successful trials.

Approval

The protocol should be formally approved by signatures representing appropriate expertise qualified to determine if the experimental plan will test the process as desired. The final signature should be the representative of the quality assurance/quality control (QA/QC) unit.

Final Validation Report

Just as the protocol presents the experimental plan to be followed, the final Validation Report presents the results of carrying out the experimental plan. There should be a nearly perfect "fit" between the protocol and the final Validation Report (i.e., the report should be organized in the same order as the protocol). The final Validation Report normally includes the following information:

- Copy of the original protocol.

- Sampling Verification Report (signed and dated by the person performing the sampling).

- Original analytical data (signed and dated by the analyst and his/her supervisor).

- Analyses and conclusions (evaluation of data, explanation of any deviations, and a statement that all acceptance criteria were met).

- Approval sheet (written, signed approval by appropriate management, including the QA/QC unit, that all conditions of the protocol were met and that the cleaning process may be considered validated.

Usually, at this point, a cleaning process becomes subject to change control. Often, the equipment and batch records are labeled as "Validated—any change requires prior approval before implementation of the change." A statement, to this effect, may also be included in the final Validation Report.

Change Control

Once a cleaning process has been validated, it should not be changed, since changes could undermine the very principles on which the validation was based, namely, that the process will be consistent and reproducible if the process remains unchanged. Some changes could obviously impact the cleaning process, such as changes in the cleaning procedure itself, the manufacturing equipment, the cleaning agents or detergents, the formulations, and the coating solutions for coated products. There are other more subtle changes, such as changes in raw materials, the formulation of the cleaning agent, the temperature of the washes or rinses, and the time the equipment remains soiled before cleaning is initiated. All of these changes or potential changes should be reviewed before the change is actually made. A written document should be prepared that describes, in detail, the proposed change. It should be approved in writing by the appropriate disciplines, including the QA/QC unit.

EMERGING TRENDS IN CLEANING IN THE PHARMACEUTICAL INDUSTRY

Many companies are experimenting with new cleaning methods and new cleaning technology in an effort to clean more efficiently and to a lower residue level. Previously, many cleaning procedures were not product specific (i.e., the same cleaning procedure was used for all products in a given facility or for all products manufactured in a certain piece of equipment). As cleaning has become more scientific, there has been a realization that the cleaning technique used must be directed toward the specific equipment and the products made in that equipment.

Some companies have modified their use of cleaning agents. There has been a recent trend away from using organic solvents for cleaning, especially in the terminal stages of the manufacturing process, because of their toxicity. Some companies have found that aqueous-based cleaning is safer than organic-solvent cleaning in terms of the potential carryover of toxic organic solvent residuals into the next product. Since organic solvents must be recovered or disposed of, they also present an environmental impact.

Some companies have begun to use more product-specific cleaning agents, while others have eliminated the use of cleaning agents in an attempt to simplify their cleaning methodology. Many of the older cleaning methods that were developed many years ago were quite arbitrary but became entrenched in plant procedures. Some companies are finding that hot water is as effective as any other agent for cleaning equipment.

Still other companies are experimenting with high-pressure water-cleaning devices. Such devices are very similar to those used in the do-it-yourself car washes, except that they may utilize considerably higher pressure. These devices may be particularly appropriate for hard-to-remove residues, such as proteins. Any user of this technique should be aware of the safety concerns that these devices present; fingers and limbs have been lost by the careless use of these high-pressure devices.

Another emerging trend is in the visualization of hard-to-access areas, such as pipes, transfer hoses, or small and intricate pieces of manufacturing equipment. The fiber optic probe allows the viewing of surfaces that cannot be accessed in any other manner.

A few companies are experimenting with using video cameras to examine equipment after cleaning. Since the cameras come equipped with a light source and can be placed on a probe unit that is indexed and covers the entire inside surfaces of large tanks, it is actually superior to having a worker climb inside. It is also safer (no fumes) and less likely to lead to further contamination from the entry process. Further, the tapes can provide visual documentation should questions arise later on.

REFERENCES

Baffi, R. 1991. A total organic carbon analysis method for validation cleaning between products in biopharmaceutical manufacturing. *J. Parenteral Science and Technology* 45:13.

Conine, D. L., B. D. Naumann, and L. H. Hecker. 1992. Setting health-based residue limits for contaminants in pharmaceuticals and medical devices. *Regulatory Toxicology and Pharmacology* 1:171.

Dourson, M. L., and J. F. Stara. 1983. Regulatory history and experimental support of uncertainty (safety) factors. *Regulatory Toxicology and Pharmacology* 3:224.

Ellsworth, D. 1996. Compliance Update—FDA. *Proceedings of the International Good Manufacturing Practices Conference,* University of Georgia, Athens, Georgia.

FDA. 1991. *Guide to inspection of bulk pharmaceutical chemicals.* Rockville, Md., USA: Food and Drug Administration, U.S. Department of Health and Human Services.

Jenkins, K. M., A. J. Vanderwielen, J. A. Armstrong, L. M. Leonard, G. P. Murphy, and N. A. Piros. 1996. *PDA Journal of Pharmaceutical Science and Technology* 50:6.

Layton, D. W., B. J. Mallon, D. H. Rosenblatt, and M. J. Small. 1987. Deriving allowable daily intakes for systemic toxicants lacking chronic toxicity data. *Regulatory Toxicology and Pharmacology* 7:96.

United States v. Barr Laboratories, Civil Action No. 92-1744, United States District Court for the District of New Jersey.

7

STABILITY AND QUALITY CONTROL

John F. Addison
David E. Wiggins

Schering-Plough HealthCare Products

A coated solid dosage form is a complex unit. In order to be successful, the concept, development, testing, scale-up, compounding, coating, validation, packaging, stability, and control areas must all mesh within a myriad of regulatory requirements. The objective of this chapter is to present the stability and control of coated solid dosage forms in a practical approach consistent with regulatory requirements. Areas that may cause problems or failures will also be discussed.

Peter Drucker said, "Long-range planning does not deal with future decisions, but with the future of present decisions." We should never lose sight of the fact that in manufacturing coated solid dosage forms, or any pharmaceutical product for that matter, we are, in effect, preparing for a present crisis but hope one never comes. To that end, the people involved in the stability and control of coated solid dosage forms must have the education and training commensurate with the responsibility for doing the job right. Doing the job right includes proper documentation. "If it is not documented, it was not done" should be the proper thought process. Aristotle once said, "Quality is not an act; it is a habit."

STABILITY

Background

The purpose of stability is to show that a drug product will meet applicable standards of identity, strength, quality, and purity at the time of use [21 CFR 211.137(a)]. This information is normally reflected on the package in the form of an expiration date and a label storage statement. The stability of coated tablets is determined much like the stability of other dosage forms. However, the unique nature of coated tablets offers some advantages and disadvantages to its stability that require specialized testing.

For a New Drug Application (NDA), stability requirements are defined by U.S. Food and Drug Administration in the *Guideline for Submitting Documentation for the Stability of Human Drugs and Biologics* (FDA 1987). In this document, the specific parameters that must be addressed include appearance, friability, hardness, color, odor, moisture, strength, and dissolution. Although a similar document does not exist for monograph over-the-counter (OTC) drug products, the requirements included in the 1987 document are basically followed by OTC manufacturers as well. The primary difference between the stability of an NDA drug product and an OTC drug product is the manner in which the expiration dating period is established. For an OTC drug product, reliance can be placed on historical, long-term stability data available for a drug substance to provide a preliminary indication of drug product stability. Or, based on short-term testing (i.e., 3 months), a 24-month expiration date may be projected. However, for NDA drugs, where a baseline of historical data is typically not available, expiration dates are established based on long-term data (i.e., 24 months or more).

In the interest of world harmonization, the FDA and the U.S. pharmaceutical industry have been working with similar groups from the European Union (EU) and Japan to establish unified stability requirements for drug products marketed in these three regions. This effort is known as the International Conference on Harmonization (ICH) of Technical Requirements for Registration of Pharmaceuticals for Human Use. Through the efforts of these six ICH members, stability requirements for drug products have been updated (FR 1994). While initially geared toward new molecular entities (NMEs), the requirements of the ICH have now been broadened to include all NDA products. It is anticipated that in the future, some aspects of the ICH requirements will also be applied to OTC drug products as well.

Requirements

Current Good Manufacturing Practices (cGMPs) require a written testing program designed to assess the stability characteristics of drug products [21 CFR 211.166(a)]. The purpose of this program is to generate data for appropriately determining storage conditions and expiration dates. The regulation mandates the program be followed and includes the following minimum requirements for coated tablets:

- Sample size and test intervals for each attribute.
- Storage conditions for samples retained for testing.
- Reliable, meaningful, and specific test methods.
- Testing the product in the same package as marketed.
- An adequate number of batches tested to determine the appropriate expiration date.
- Accelerated studies to support tentative expiration dates.
- Confirmation of all tentative dates.

On May 3, 1996, the FDA published a proposed rule (Docket No. 95N-0362) in the *Federal Register* to amend certain requirements for finished pharmaceuticals. This proposed rule included a change in the current stability testing requirement to add at least one batch of each drug product to the stability program annually. This is done to provide ongoing stability after the expiration date has been determined. Most companies were doing this already.

The first three batches are placed under test in the stability program to account for variability and confirm the expiration date previously estimated. Subsequent batches are included to account for changes that may occur (i.e., personnel, raw materials, suppliers, equipment and manufacturing environment). Any of these changes may impact product stability.

Although the regulatory requirements for stability seem complete, companies may need to add extra test parameters and generate pertinent additional data and information to better assess how the product will perform in the field. While the cGMP requirements must be met to market the product, the company should know the product characteristics and performance beyond the required tests. Examples of additional tests to generate additional data for this assessment may include freeze-thaw cycles, 50°C stress, ultraviolet (UV) and fluorescent light testing, and simulated shipping tests. It is the responsibility of the company to prepare and implement a well-thought-out stability testing plan to provide this product assessment prior to marketing.

Packaging

Probably the first aspect to consider before assessing the stability of a coated tablet is what type of package (container/closure) will be used to market the product. For tablets, choices could include bottles, blisters, and pouches, and the manner in which the product will be marketed (prescription vs. OTC) may dictate the type of package required. The fact that coated tablets provide a special barrier to moisture means that less expensive options (e.g., polyvinyl chloride [PVC] vs. Alcar®) may be used. However, some types of coated tablets may be susceptible to both moisture and light and, thus, require highly protective packaging. If the intent is to market multiple package types, then all package types should be included in the stability program. Obviously, the packages used for stability must be the same as those that are intended to be marketed.

The U.S. Pharmacopeia (USP) requires some pharmaceutical dosage form packages to limit levels of moisture-vapor transmission (i.e., a tight or well-closed container). If these packages are specified, they must pass the tests required in the USP. Caution should be exerted when using bottles containing cotton for products with moisture-vapor requirements. If the packaging line is not adjusted properly, the cotton is not placed down into the bottle, or the knives are dull, a strand (only one small fiber) may overlap the lip of the finish and serve as a conduit for moisture that may contribute to stability failure. It is much easier to assure this does not happen before testing than to explain bad results later or cause long delays in getting to market with the product.

Parameters

The product attributes that must be addressed in the stability program for a tablet were highlighted above and, for a coated tablet, all but hardness should be included. For coated tablets, special consideration must be made for each of these parameters, since the way that a coated tablet ages on stability will be different from an uncoated tablet.

For the appearance parameter, coated tablets have two additional attributes that must be monitored. Coated tablets are typically polished, and any signs of a loss of luster must be included in an examination for stability. Signs of changes in the coating quality (splotching, chipping, cracking, etc.) must be monitored. This aspect is particularly important for enteric coated or extended-release tablets, where this type of change may lead to other product attribute failures (i.e., dissolution). While these type changes may be considered subjective and not as serious as an assay failure, they are very important since the appearance of a coated tablet is a consumer's first impression of the quality of that tablet.

The friability parameter (Tablet Friability, USP <1216>) presents an additional challenge in developing a coated tablet, since the coating, particularly for sugar-coated tablets, makes the tablets much harder. Unlike uncoated tablets, where friability may lead to powdering of a tablet, friability of a coated tablet may lead to total tablet breakage. Again, for an enteric or extended-release tablet, this could result in other parameter failures.

Because of the polish on coated tablets, color is a much more important parameter than on uncoated tablets. Colors are typically brighter, and fading is much more likely to be a problem. Tablets must be exposed to both UV and visible light in order to assess color stability, since clear blister packaging exposes the product to light. Color changes are usually determined visually and made by comparison to a control (same lot unexposed to light conditions). Color changes on stability may require special UV–absorbing blister material or may prevent the use of blisters altogether.

Odor is typically not perceptible for coated tablets, unless residual coating solvents are not properly removed during the manufacturing process. However, some coated tablets can occasionally develop a musty odor; thus, odor should be monitored on stability.

Moisture uptake by a tablet is one of the most common problems that leads to active ingredient degradation and physical attribute failures in a tablet. As such, the moisture determination of tablets is a very important attribute to monitor in a stability program. Moisture content can be determined by using either water titration methods (Water Determination, USP <921>) or loss on drying methods (Loss on Drying, USP <731>). For the analyst, coated tablets present an interesting challenge in moisture determination, since the coating itself is designed to prevent moisture penetration into the core of the tablet. For coated tablets, methods to determine surface moisture uptake are appropriate, even though they are more difficult to perform, since the surface moisture of a coated tablet could affect the dissolution profile.

The active ingredient content of a coated tablet must be monitored, since strength is directly correlated to effectiveness. Assays that are accurate, sensitive, specific, and reproducible [21 CFR 211.165(e)] for each active ingredient are required. These methods can be found in many different compendia or may be developed in-house by the analytical group. Inadequate grinding of coated tablets prior to analysis is one aspect that makes coated tablets more difficult to analyze and can lead to erroneous results. As for uncoated tablets, degradants must also be identified and monitored throughout the stability testing of coated tablets.

Dissolution of coated tablets as defined in the USP (Dissolution, <711>) presents the biggest challenge for pharmaceutical scientists developing a coated tablet. Since most coated tablets are either modified release, extended-release or enteric coated, dissolution is the most critical factor in determining product stability. Dissolution of coated tablets can be affected by many variables, some of which are even in the analysis of the tablets. Many FDA recalls are due to the product not meeting dissolution specifications, and coated tablets are no exception. This is particularly true because timed-release dosage forms have multiple time points that must be monitored.

Including all of these parameters in the stability program for a coated tablet will enable the pharmaceutical scientist to adequately determine the stability of the product. It is extremely important to reiterate here that the personnel involved in generating the chemical and physical data in the stability program be experienced in all aspects of the functions they perform. Not reporting even a slight physical anomaly can manifest itself as a large problem with time. Therefore, the personnel involved should be trained to express what they observe accurately, without conclusions, until that testing interval has been completed and reviewed by supervising personnel, including the formulation chemist. Microbiological parameters will be discussed later in this chapter.

Storage Conditions

Storage of a drug product on stability falls into three categories: long term, accelerated, and stress conditions. Each condition has an important role in establishing the stability profile of a drug product. Not all are required, but prudent planning will at least address all three.

The drug product in its marketed container/closure system will be stored at some primary room temperature condition. For most tablets, this would mean long-term storage at ambient room temperature, although some less stable drug substances could require storage under refrigerated conditions. The ongoing debate between the FDA and the pharmaceutical industry has been over how to define room temperature conditions. While everyone recognizes that, in the real world, a marketed drug product will not always be stored in a tightly controlled environment, the debate has been over what isothermal conditions (25°C or 30°C) should be used to try to mimic this variable environment to which a product will be exposed. Consensus among the ICH participants is that this condition is 25° ± 2°C and 60 ± 5 percent relative humidity (RH). For non-NDA OTC products, 3 months of acceptable stability data under 40°C and 75 percent RH will allow the granting of a tentative 24-month expiration dating period. This is often referred to as the Joel Davis test, from the concept presented by Joel Davis of the FDA to the pharmaceutical industry at a manufacturing controls seminar of The Proprietary Association, now known as the Nonprescription Drug Manufacturers Association, in 1978.

For NDA products, the same accelerated conditions are also necessary, but 6 months instead of 3 months of data are required. Under the ICH, an alternate condition (30° ± 2°C and 60 ± 5 percent RH) can be used if product failure occurs at 40° ± 2°C and 75 ± 5 percent RH, but it must be followed for 12 months. Unlike OTC products, the accelerated condition for NDA products does not allow any expiration dating projections; rather, it is used to establish a label storage statement.

Since the drug product may occasionally be exposed to extreme conditions of temperature, light, or humidity, stress conditions should also be included in the stability program. For coated tablets, low temperatures (e.g., 4°C) are usually not a problem unless they lead to additional moisture uptake by the tablets. High temperatures (e.g., 50°C) can prove to be more of a problem and can lead to physical changes, such as cracking or splotching. Light also can adversely affect the appearance of tablets, and exposure to intense UV and visible light can mimic long-term exposure to light. Data generated from stress conditions may not necessarily result in a formulation failure but rather alert the pharmaceutical scientist to potential product change if exposed to such conditions. Stress condition test failures usually warrant label warnings.

As defined in the cGMPs, an expiration dating period is related to specific storage conditions [21CFR 211.137(b)], and these storage conditions are defined in a label storage statement. Typically, this statement, which appears on the label, indicates the condition under which the product should be stored throughout its expiration period. Unfortunately, label storage statements are not consistently used throughout the industry, even though they primarily define controlled room temperature (25°C). For coated tablets, additional warnings may be included that specify protection from light or humidity when it is found that these stress conditions adversely affect the product.

Timing

Products stored under various conditions are typically tested at 3-month intervals the first year, 6-month intervals the second year, and annually thereafter. A stability protocol that defines the tests that will be run, the conditions of storage, and the length of storage must be prepared before any studies are initiated. For coated tablets, the protocol should include all of the parameters cited previously, along with the three types of storage conditions.

For NDA products, 12 months of real-time data are required prior to submission of an NDA to the FDA. Thus, stability studies must be initiated early enough to allow timely product introduction. Once 18 months of data have been generated, the FDA expects to receive an update of the stability data.

For OTC products, typically 3 months of room temperature, stress, and accelerated condition stability data are generated prior to entering the marketplace. However, since long-term real-time data do not exist for these OTC products, the cGMPs require that ongoing studies be conducted to confirm the tentative expiration dating period assigned [21 CFR 211.166(b)].

Expiration Projection

The expiration dating period of an NDA product is based on a statistical analysis of long-term room temperature data. The statistical model allowed by the FDA is defined in their 1987 stability guideline and is based on an article published by Cartensen and Nelson (1976). Projections beyond actual data are allowed by the FDA if the statistical analysis of the data supports the projection. However, extrapolations of more than 6 months beyond the real-time data available are rare.

It is possible to use the same NDA approach to establish the expiration dating period for OTC products. However, because the time frames are so long, this method is rarely used by OTC drug firms. The Joel Davis test is typically used to establish the expiration dating period of OTC drug products. A 24-month expiration dating period can be assigned once 3 months of acceptable accelerated stability data have been generated. A 36-month expiration dating period can be assigned if 4.5 months of acceptable accelerated testing are completed. However, this latter approach is more risky, since a 36-month expiration dating period is assigned based on only 4.5 months of data. Alternatively, real-time aging can be combined with a repeat of Joel Davis accelerated testing to extend the expiration dating period beyond 24 months. This is accomplished by adding the 24-month projection from the accelerated testing to the age of the tablets at the time the study was initiated (i.e., 12-month-old tablets subjected to 3 additional months of accelerated conditions).

An earlier and somewhat more theoretical approach to projecting expiration dating periods involves the use of Arrhenius plotting. This technique relies on the relationship between a reaction rate constant (k) and temperature (Kelvin). Although a detailed discussion of Arrhenius plotting is beyond the scope of this chapter, the technique basically involves performing assays with a highly accurate method on samples that have been stored at several different temperatures (35°C, 45°C, 55°C) over time (0, 1, 2, 3, 4, 5, 6 months). These data are then used to calculate the stability rate constant, which in turn can be used to calculate an expiration projection. This approach is cumbersome, labor intensive, and offers no advantages over other techniques. It is also not approved for use by the FDA and is not widely used within the pharmaceutical industry.

It remains good policy for a key person in the company filing the NDA to develop a rapport with the FDA reviewing chemist. It is always easier to establish and obtain agreement with the FDA on the front end of a study or change rather than risk repeating lengthy studies or experience nonapproval by the FDA.

Stability Commitment

After labeling issues, the next most common reason for an FDA product recall is stability failure. The discussion up to this point has been primarily research in nature, that is, the stability requirements necessary before going to market. However, the cGMPs require that production batches be added to the stability program on an ongoing basis. If a product failure occurs on production stability batches, a potential recall situation exists.

Because of this potential liability, stability scientists must thoroughly evaluate the stability profile of all new dosage forms when establishing the expiration dating period of a new product. Coated tablets present an additional challenge since physical changes in the coating are hard to predict, even though they may dramatically affect the release rate of the drug from the tablet. Careful evaluation during the scale-up of coated tablets can reduce the risk, but the potential for changes over time is always a possibility.

PRODUCT REVISIONS

After a product is developed and placed on the market, changes may be required in the raw materials, formulation, processing, or packaging for any number of reasons. Many of these changes will require additional stability data. For NDA products, the FDA has begun to issue a set of guidance documents, referred to as Scale-Up and Postapproval Changes (SUPAC), that define what must be done before a change can be implemented. For OTC products, application of judicious scientific judgment is required to assess the impact of the change. In either case, before any changes are made, the effect of the change on the stability of the product must be addressed.

Raw material vendor changes are one of the most common types of postmarket changes that occur and are often the most troublesome. While changing the vendor for a raw material may seem innocuous, small changes in the synthetic pathway or impurity profiles can have an adverse effect on the stability of the coated tablet. Depending on the raw material, some amount of stability data is typically warranted prior to initiating the change on marketed batches.

Most formulation changes of coated tablets will affect the stability of the product, and repeating the original stability program may be necessary. For minor changes, only limited stability data may be required. Typical formulation changes are addressed in the SUPAC documents, and these requirements must be considered for NDA products. These requirements may also be used as a guide for OTC products.

A need to revise the manufacturing process may occur during the life of a coated tablet due to technology changes. Because these process changes will likely affect how the product is coated, stability studies will need to be repeated. However, since no chemical changes are being made to the formulation, performance parameters (e.g., dissolution, assay) and physical parameters may be the only parameters that need to be monitored on the revised product.

For packaging changes, the effect on the package barrier properties should be evaluated before the decision is made to redo stability. For changes that are external to the product contact surface, such as adding a child resistant feature, no stability data will be required. However, even though changing from a blister to a bottle offers advantages in terms of less moisture penetration and more product protection, increases in tablet breakage may occur. Package size changes usually do not require stability data, but changing to a totally different package type will require stability data.

The amount of work involved to make a product or package revision can be rather intensive. Before the decision is made to make a change, the overall cost must be assessed and weighed against the advantages of making the change. By doing this type of analysis, it may be determined that a change that may benefit one group within a pharmaceutical operation does not outweigh the time and cost required to implement the change. Again, sound scientific judgment must be used in making these decisions.

QUALITY CONTROL

This book is about coating solid dosage forms, but the control of such must cover a much broader area than just coating. Someone once said that "insanity is doing things the same way every time and expecting a different result." Each process is inherently different. All coating processes are not alike, and the controls applied to them should have specific attributes. This section of this chapter will focus on a product life-cycle approach to the control of coating solid dosage forms.

Background

During 1996, the FDA "Enforcement Reports" listed almost 100 recalls involving solid dosage forms. Approximately 90 percent were prescription and 10 percent were OTC products. The results were interesting.

Approximately 50 percent of the recalls involving solid dosage forms were for potency. Except for cGMP deficiencies, one could only guess what happened during the manufacturing process. The point here is that many things can happen during the manufacturing process that may cause problems, even a product recall. It would be a waste of valuable time to have a high-quality coating process only to have the tablets recalled due to a problem that happened before or after coating. Good omelets do not come from bad eggs.

Among the reasons given for the recalls were the following:

- Does not meet content uniformity specifications.

- Subpotent or superpotent.

- Potency not ensured by the expiration date.
- Stability test failure.
- Failed dissolution prior to expiration date.
- Color change.
- Metal particles or wire or in the tablets.
- Wrong ingredient added.
- Mislabeled shipping cases.
- Label and carton directions do not agree or label printing error.
- Obsolete package insert.
- Wrong expiration date.
- Different strengths of same product mixed or wrong tablets in bottle.
- cGMP deficiencies.

Requirements

The responsibilities of the quality control (QC) unit are clearly defined in 21 CFR 211.22a–d. On May 3, 1996, in the proposed rule to amend the cGMPs (Docket No. 95N-0362), the FDA further clarified these requirements by making the QC unit responsible for keeping validation current to ensure that compliance is more consistent and reliable. This great responsibility requires an adequate number of highly trained and experienced staff to help keep products off the recall list.

Development

It is in the design stage where quality begins. Quality cannot be tested into a product; it cannot begin on the production floor; it must begin in the research and development (R&D) laboratory. When quality is not introduced in the design phase, many negatives can occur, such as lost manufacturing time, retesting, investigations, rework, delayed shipments, and more.

The Design

The design includes defining the process endpoint and how this endpoint will be measured. The process is selected to best achieve this endpoint and verified that it does accomplish the design, as well as a proposed range of acceptable values for the product. The process must be able to sustain these acceptable values on a continuing basis.

The Product Profile

The product profile will include the business and technical aspects of the product. The business aspect addresses, for example, sales, cost of goods, price, and coated solid dosage product. The technical profile includes shape, single or double layer of coating, printing on the tablet, and the amount of stability data required to go to market, whether or not it is a USP product or requires an NDA/ANDA (Abbreviated New Drug Application). Concurrently, packaging configuration such as bottle (and shape) or blister, tablet count, carton, and label are profiled. Tamper evidency and child-resistant requirements should also be determined during this phase of development.

The Plan

Using the information developed during the design and product profile stage of development, the actual development of the product takes place. During this process, certain outputs are expected. The final process equipment is selected, along with primary packaging specifications. Manufacturing directions and formulas are developed. Tentative product specifications are also established, including the following:

- *Thickness:* QC testing with a micrometer or automated means. Thickness specifications should be synchronized with hardness, weight, friability, and uniformity.

- *Microbiology:* Tests to determine and control the potential for microbiological contamination (discussed later in the chapter).

- *Organoleptic:* Visual appearance and characteristic odor determination performed by comparing the product being tested with a reference standard of acceptable organoleptic quality. This test is performed with the eyes and nose and is extremely valuable for detecting anomalies that, with instrumentation, often go undetected until the consumer uses the product.

- *Potency:* Quantitative determination of active ingredient(s) by a compendial or validated analytical method. Usually reported as percent of theory.

- *Dissolution:* Bioavailability and QC testing that can measure formulation factors when performed according to USP 23 <711>.

- *Content uniformity:* Performed to measure the uniformity of dosage units when done according to USP 23 <905>.

- *Hardness:* Measures tablet or core hardness. Strong-Cobb hardness tester or equivalent may be used.

- *Friability:* Measures the ability of tablets to withstand shipment without breakage when performed in accordance with USP 23 <1216>.

The equipment used in the above tests is available from laboratory supply houses, such as Vankel Industries, Inc.[1], or Erweka Instruments, Inc.[2]

Once methods development, solubility profiles, cleaning methods development, and process workup are completed, preliminary chemical and physical stability studies are expected, along with formula rationale and manufacturing confidence. During this important time, methods validation needs to be completed. This includes stability versus USP, cleaning methods, and routine QC methods.

Changes are expected and acceptable during the development process, provided they are properly documented and justified. QC unit personnel should be heavily involved in the process development stage and thereafter. A thorough understanding of the formula, process, specifications, and equipment is necessary to minimize problems and help with timely decisions. When the scale-up batches are completed, all systems should be ready for process validation batches.

Process Validation

Process validation is establishing documented evidence that provides a high degree of assurance that a specific process will consistently produce a product meeting its predetermined specification and quality characteristics. Prior to starting the process validation run, a detailed protocol is necessary, clearly stating every objective of the validation study. This protocol includes process monitoring and sampling and testing details.

---

1. Vankel Industries, Inc., 36 Meridian Road, Edison, NJ 08820.
2. Erweka Instruments, Inc., 56 Quirk Road, Milford, CT 06460.

The details of process validation will not be covered in this chapter. However, the validation process is referenced because many important aspects of producing quality solid dosage forms are defined during that process.

Control

21 CFR 211.110 states,

> . . . Control procedures shall be established to monitor the output and validate the performance of those manufacturing processes that may be responsible for causing variability in the characteristics of in-process material and the drug product.

While the primary objective of this chapter is focused on the coating of solid dosage forms, an adequate presentation of the control of the process must include aspects of the preparation of the entire dosage form. Therefore, the following will reflect this fact, with a focus on tablets.

These control procedures are far reaching and include not only the various manufacturing processes, but plumbing, lighting, the HVAC (heating, ventilation, and air-conditioning) system, washing and toilet facilities, sanitation, personnel, and maintenance. Should one or more of these areas drift from the expected or malfunction during the process, the product being manufactured at the time may be compromised. An effective monitoring program to ensure compliance in these areas increases confidence in the process and minimizes variability. Simple checklists work well as a tool to monitor these areas.

Personnel

21 CFR 211.25 states,

> Each person engaged in the manufacture, processing, or holding of a drug product shall have education, training, and experience, or any combination thereof, to enable that person to perform the assigned functions.

These requirements continue and are the same for supervisors. It is also made clear that an adequate number of these qualified people are necessary to do the work. Training in cGMPs is required on a continuing basis and should be conducted by people who know what they are doing. All training must be documented.

During the processing of validation batches, the presence of large numbers of highly technical and knowledgeable people from R&D and QC will be apparent. When these validation runs are completed, these people return to their offices and laboratories to work on other important projects, abruptly turning over the manufacturing process to manufacturing personnel and QC inspectors or auditors. Either of two things is likely to happen. First, sales of the product will fall and manufacturing of the product will not be necessary beyond one shift, or the product will be made once or twice during the year to meet sales requirements. Second, sales will grow at an accelerated pace and extended hours will be required in manufacturing to meet the sales need. These additional requirements may include double shifts or weekend work. In either scenario, new people may be required for the manufacturing process who did not experience the validation runs; therefore, new training is required to ensure the best efficiency and compliance. Infrequency of manufacturing creates the same type of challenge. Personnel turnover, vacations, leave, and illness also may contribute to variations in personnel. These challenges are real, and mechanisms must be in place to deal with them effectively or batches may be lost, at great cost to the company.

Raw Materials

The raw material sources (suppliers) and specifications are developed, tested, and set during product development. QC verifies that incoming raw materials conform to the specification requirements. If it were that simple, no raw material problems would ever occur.

Suppliers should be visited and formally approved by the QC unit prior to manufacturing. For various reasons, this is not always done. For example, a supplier may actually prove to be a distributor that buys from many manufacturers based on price. This can result in significant quality variations from batch to batch of raw materials. Therefore, on receipt of raw materials, the approved supplier should be confirmed.

Chemical and physical testing of all raw material shipments is a cGMP requirement. Physical requirements are equally important. Particle size is always important, but especially in dry granulations for direct compression.

Taking representative samples from each shipment is a cGMP requirement, although a specific plan is not given. A typical sampling plan may include taking a sample from each drum if the shipment contains five or less containers. If the shipment is larger than five containers, then sample randomly the square root plus one amount of containers. Depending on history and experience, it may be necessary to sample from the top, middle, and bottom of the container. Bulk shipments of raw materials, such as totes, tank wagons, or rail cars can and should always be sampled from the top and bottom.

It is important to store raw materials in a manner consistent with the manufacturer's instructions. Although the intent may be to use all of the raw materials each time, or at least a unit container each time, it almost never works out that way, necessitating lengthy storage that may turn into several years. Many things can happen during raw material storage that may impact the manufacturing process, such as increased moisture, caking, deterioration, and other undesirable changes. In order to minimize problems, a written program should require the retesting of raw materials for compliance at specified intervals, such as within 6 or 12 months prior to use. Such a program can save a lot of grief and cost.

Raw materials should be screened or filtered at the point of manufacture. This step will avoid problems such as lumps and tramp debris. All raw materials should be passed over a strong magnet at the point of manufacture to remove tramp metal particles from the process. It is sometimes amazing to see what actually comes along with some raw materials. The microbiological aspects of raw materials will be discussed later in this chapter.

In-Process

21 CFR 211.110 states,

> To assure batch uniformity and integrity of drug products, written procedures shall be established and followed that describe the in-process controls, and tests, or examinations to be conducted on appropriate samples of in-process materials of each batch.

A flowchart (Figure 7.1) of the process should exist and be used for determining critical monitoring points. This flowchart should be finalized by the time validation batches are produced. The setpoints of the equipment should be monitored, along with adherence to manufacturing instructions exactly as listed on the batch formula sheet. The cleanliness of the equipment and critical room operating parameters should be logged, such as temperature, humidity, particulate levels, and bioburden. It is obvious that mix times, rates of addition, and the order of addition are important. If equipment utilizing screens is used, such as oscillator granulators, the screen should be verified as intact at the beginning and end of the formulation step to ensure no breakage has occurred and no fragments

Figure 7.1. Flowchart of a process for determining critical monitoring points.

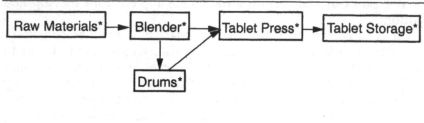

*Control points in the process

have fallen into the granulation mixture. The importance of specific written directions and specifications cannot be overemphasized. For example, the failure to specify and monitor the amount of solution used for granulation may result in dissolution failures in batches of product, causing regulatory problems or even product recall.

Blenders can be a problem source—dead spots that create content uniformity failures. High-shear or high-energy mixers can generate heat high enough to char some ingredients. For wet granulations, it is important to control the rate and amount of solvent addition because some of the drug substance may partially dissolve and recrystallize in a different physical form. Also, many of these mixers cannot be cleaned properly without disassembling the intensifier bar. Cross-contamination may occur if blenders and mixers are not cleaned properly.

Dryers can also be a source of problems. Cross-contamination can occur when bag filters are not dedicated. Dryers should be validated as to heat uniformity to ensure consistent moisture in granulations at the end of the drying cycle.

The potential for problems to happen exists at any time during the manufacturing process, especially when sophisticated equipment is used. These areas of concern should be addressed during product development, with problem areas identified and confirmation that the system is in control during the validation runs.

Blender sampling and testing is an important part of any process monitoring program. When manufactured in accordance with formula directions, Table 7.1, which indicates sampling points, specifications, and batch identification, can help to assess product quality. Unit dose sample sizes are taken at specified points in the blender and tested according to specifications.

Table 7.1. Blender Sampling and Testing

Blender Sample Location	Specifications	Batch #
Top left	85–115% of theory	
Top right	85–115% of theory	
Middle left	85–115% of theory	
Middle right	85–115% of theory	
Bottom	85–115% of theory	
Average		
% relative standard deviation (RSD)	5% or less	

The same sample approach can be done for drums or totes. Table 7.2 can be used. Drum or tote samples from preselected places in the batch are submitted to the control laboratory for particle size analysis, bulk density, and angle of repose testing. The results can be tabulated as in Table 7.3. Powder fineness is found in USP 23 <811>.

Table 7.2. Sampling of Drums or Totes

Drum Sample Location	Specifications	Batch #
#1, Top	85–115% of theory	
#1, Middle	85–115% of theory	
#1, Bottom	85–115% of theory	
#3, Top	85–115% of theory	
#3, Middle	85–115% of theory	
#3, Bottom	85–115% of theory	
#6, Top	85–115% of theory	
#6, Middle	85–115% of theory	
#6, Bottom	85–115% of theory	
Average		
% RSD	5% or less	

Table 7.3. Particle Size Analysis

Particle Size	Specifications	Batch #
20 Mesh %		
40 Mesh %		
60 Mesh %		
80 Mesh %		
100 Mesh %		
Bulk Density (g/mL)		
Angle of Repose		

After the validation runs and enough additional batches have been produced to confirm statistically that the process is under control, use the data to justify and validate that reduced sampling and testing may be in order.

Compression

Granulations with good flow characteristics and good uniformity tend to cause fewer problems during the compression process.

The monitoring program should include a general inspection of the pressing room prior to batch processing to ensure that the room is in order and suitable for use. Particular attention should be paid to the press, that it has been properly prepared for hopper feed, feed frame flow, die fill, ejection, press speed, and ease of compression. Confirmation should determine that the compression limits are within specifications and the physical appearance of the tablets are satisfactory. Initial checks of hardness, thickness, and weight are also made.

Tablet weight variation testing should have been performed during the process validation stage. If this was not done, it needs to be done. Tablet weight variation should be done on samples taken from the front and back of the press at the beginning, middle, and end of the run. In addition, samples should be run bracketing the speeds of the tablet presses to be used.

Data should be grouped showing frequency versus percent of average tablet weight at the different press speeds. Process capabilities should be calculated to confirm tablet weights at each speed. These process capabilities will reflect the ability of the process to meet the weight variation specification of the tablets during compression.

Sample plans vary depending on many factors. A typical sampling plan may include sampling from the beginning, middle, and end of each shift. Testing will include organoleptic, hardness, thickness, friability, tablet weight, content uniformity, disintegration, and assay. The results may be tabulated as in the example in Table 7.4.

Samples are pulled from each station. Sampling can also be pulled by operational hours, which is defined as 60 continuous or cumulative minutes of pressing. Any deviation from the written sampling plan must be approved by the department manager or area supervisor; justification must be documented on the Batch Summary Report or in the batch records. If any sample fails to meet test limits, area supervision should be consulted immediately for corrective action and disposition.

After a significant number of batches have been produced, subsequent to the validation runs, the data may warrant reduced

Table 7.4. Results from Sampling (Example)

Test	Specifications	Time	Station	Batch Results
Organoleptic	comparable to standard			
Hardness	5.0–9.0 SCU (Strong Cobb Units)			
Thickness	0.220–0.230 in.			
% Friability	not more than 1.0%			
Content uniformity range				
Content uniformity % RSD	not more than 5%			
Tablet weight, mg	325–335 mg			
Disintegration	not more than 30 sec			
Assay, mg	180–220 mg			

sampling. Care must be exercised here and sign off obtained from R&D, QC, and manufacturing. Should this product be manufactured under a NDA, the regulatory department should also approve the change.

Coating

Prior to beginning the coating process, a general inspection of the area should be completed and documented. The cleanliness of the equipment is checked, along with other critical parameters (e.g., room temperature, humidity, airflow, and equipment setpoints). Assurance is obtained that coating solutions and tablets to be coated have been approved by QC.

After the coating process is complete, samples are taken from each coating pan for physical and chemical testing. The tests to be performed are similar to those done on pressed tablets and include organoleptic, physical appearance, hardness, thickness, friability, assay, dissolution profile, content uniformity, and total degradation products, if required.

The physical appearance of coated tablets gives a good indication about the quality of the coating process. A good monitoring program will always include a thorough examination of physical appearance. The classification of defects has always been a debatable item for many companies. Effective communication is necessary when developing defect classifications for tablets. Some defects are obvious, such as incomplete coating. Specifications must be synchronized throughout the process. For example, the thickness specification in sugar-coated tablets should be coordinated with the packaging department to ensure that the tablets will fit the blister packs. Thickness variation during production may require sorting for size prior to packaging to ensure efficiency. A good policy to implement on defect classification is to provide a "defect board" (listing all potential defects) in the manufacturing area for all to observe. Actual tablet samples peculiar to the tablets being produced may be included on the defect board. In the absence of actual defect samples, pictures are the next best method of consistently administering defect classifications. Branding defects are best considered separately.

Charts and tables made to record and accumulate data should be concise and simple. An example of a process QC defect and audit report is given in Table 7.5. This chart is prepared for a sample size of 200 tablets, with a noninclusive listing of some possible defects. Each defect list should be tailored to the specific tablet

Table 7.5. Process QC Defect and Audit Report

Product _____ RB# _____ Batch _____ Date _____

Critical Defects Accept 0 Reject 1	Sample #1 200 Tablets	Sample #2 200 Tablets
Defect type Foreign tablets Extraneous substance Other Total critical		
Major Defects **Accept 7 Reject 8** Malformed Incomplete coating Broken Chipping Cavitation Surface Blemish Other Total Major		
Minor Defects **Accept 12 Reject 13** Mottling Raised surface Chipping Cavitation Dull coating Dark spots Other Total Minor		
Branding Defects **Accept 8 Reject 9** Illegible brand Heavy offset ink Missing brand		

Continued on next page.

Continued from previous page.

	Sample #1 200 Tablets	Sample #2 200 Tablets
Other Total Branding		
Disposition	**Pass/Fail**	**Pass/Fail**
Comments:_____		
_____		
_____		

Inspected by _____   Reviewed by _____

Date _____   Date _____

being coated (i.e., oval tablets would not have a defect listing of chipped corners).

Definitions of defect categories should be clearly stated, and employees should be trained to understand the interpretation. Critical, major, and minor defects may be defined as follows.

- *Critical defect:* any defect that would result in unsafe or hazardous conditions for the consumer or the company.

- *Major defect:* an irregularity of a quality characteristic that may significantly reduce the elegance of the product, leading to negative impact on customer satisfaction or materially reducing the usability of the product for its intended purpose.

- *Minor defect:* an irregularity of a quality characteristic that does not affect the elegance of the product to such an extent as to reduce the usability of the product for its intended purpose.

- *Branding defect:* brand not legible or the condition of the brand significantly reduces the elegance of the product, leading to a negative impact on customer satisfaction.

The samples are examined using a suitable device for viewing both sides of the tablets. Orient the branding to ensure that the brand was applied to each tablet. Determine the severity of any defect found by comparing the defect to the classification standards previously approved. If questions arise or an unencountered de-fect is noted, consult a QC supervisor for clarification. The supervisor should also be consulted if any sample fails to meet specified limits.

Testing

The establishment of specifications, sampling plans, standards, test procedures, or any other control mechanism is developed by the appropriate organizational unit and approved by the QC unit. Any deviation from these written policies and procedures must be recorded and justified. All instruments, gauges, and equipment must be calibrated according to written procedures.

The containers and closures for sampling should be specified (e.g., metal, glass, or plastic), as well as if they should be sterile or nonsterile. The sample size is also specified, along with written details of sample collection and any precautions to be taken.

Generally, tests such as hardness, thickness, friability, and tablet weight are performed in the in-process QC laboratory, while the assay and dissolution tests are performed in the main QC laboratory equipped for the more sophisticated analytical tests involved. Assays for potency are straightforward, as the methods used are usually compendial and validated under conditions of use. Dissolution testing, at least early on, should be done as a dissolution profile and percent dissolution should be plotted versus dissolution time. Typical charts for dissolution profile data can be recorded in tables similar to Table 7.6. Times may vary according to the product being tested.

The validation batches should have dissolution profiles comparable to the biobatches to indicate equivalency of the

Table 7.6. Dissolution Profile Data

Time	Vessel 1	Vessel 2	Vessel 3	Vessel 4	Vessel 5	Vessel 6	Average
5 min							
10 min							
15 min							
20 min							
30 min							

manufacturing process. FDA investigators will surely review this important testing, along with many other important areas.

Any product failures are very serious and require full investigation and identification of an assignable cause. Product failures should be reviewed by management and a disposition made.

Microbiological

Regulatory

There are few, if any, tablets, coated or otherwise, that need to be sterile. However, they should be commercially sterile. Commercial sterility is a term from the food industry that encompasses the microbiological attributes for nonsterile food products found in USP 23 <1111>. To paraphrase these requirements, the finished dosage form does not contain microorganisms of a number or a kind likely to cause harm to the product or user. The European regulatory agencies have been requiring microbiological testing for solid dosage forms, including tablets, for a number of years. Recently, specific numerical requirements have begun to appear. Some of these requirements are as low as 2 colony forming units (CFU) per tablet or the absence of any detectable microorganisms. The USP does not specifically state that such testing is necessary to show compliance with the general microbial content requirements for nonsterile drug products.

Practical Considerations

A tablet containing viable microorganisms poses the same potential hazard to the user as any other oral dosage form on a weight/weight basis. However, the weight of a tablet taken per dose is normally far less than an equivalent dose of a liquid, thus minimizing the potential number of ingested microbial contaminants. Tablets also have several physical characteristics that minimize the risk of microbial contamination, in particular, the amount of moisture. Microorganisms require a minimum amount of available water for growth and survival. This value is referred to as the water activity level, a ratio between bound and unbound water. A value of 0.80 or above indicates a high risk for the presence of viable bacteria. Most tablets have a water activity level below 0.40. Excessive moisture usually results in distinct physical changes to tablets, which affects hardness, friability, thickness, or other such values. These factors have historically been the rationale used by most tablet producers to achieve a minimum or no microbiological level in the testing program for tablets.

In recent years, there have been several recalls for microbially contaminated tablets. In one recall of an uncoated antacid tablet, visible mold was seen on the tablets in a blister pack and in bottles. This illustrates a potential problem with tablets—the ability to absorb moisture on the tablet surface and thus create conditions suitable for microbial growth and survival. Another concern is the increasingly conservative attitude in various regulatory agencies concerning the potential harm from very low-level microbial populations. The basis of this position is that drugs are administered to individuals with potentially reduced resistance to infection because of compromised immune systems. In practical terms, it would be very difficult to refute this position successfully.

Control of Microbial Levels

The microbiological control of tablets is achieved in the same manner as for any other dosage form. This involves raw materials, process parameters, storage conditions, and chemical preservatives. In the food industry, this process has been formalized and is called a HACCP (Hazard Analysis and Critical Control Points) program.

In theory, the maximum number of microorganisms in a finished dosage form is the total contributed by the raw materials and manufacturing equipment divided by the weight of the tablet. This assumes that at no point in the process are conditions suitable for microbial growth. As a general rule, the manufacturer should be more concerned with preventing the development of a microbial population than reducing it to a specified level. Because of the many uncertainties in reliably reducing a microbial population, a more effective program should focus on prevention. Metabolic by-products of the contaminants may have multiple adverse effects on the solid dosage form and its coating, including off odors, tastes, pH shifts, degradation of actives, and so on. Another concern is the potential for contaminating manufacturing equipment, weighing areas, environmental systems, or personnel.

There are various activities involved in tablet manufacturing. Some of these activities can contribute to or reduce microbial levels. Others are neutral in their effect on microbial populations. The following sections address these activities and their potential for microbial effects.

Raw Materials

Raw materials are the first control point. It is important to have microbial specifications developed around the process and the specifications for the finished tablet. Certain raw materials, such as some dyes, sugars, and water, are particularly prone to having microbial contaminants. A good approach is to assume a raw material is a real or potential microbial problem until shown otherwise by a valid evaluation process. Commercial coating solutions are particularly prone to microbial contamination. In these solutions, antimicrobial preservatives are usually not stable or effective. It is not unusual to see the development of high populations of mold or bacteria within weeks or months of manufacture.

Bulk-stored materials (e.g., sugar, starch, and similar items) need to be on a continuous monitoring program, with particular attention to storage silos and transfer systems.

Equipment

Equipment can and will contribute to the bioburden of materials manufactured. Cleaning and sanitization procedures are the common control methods. Emphasis should be placed on the condition of the equipment when used. A common mistake is to assume equipment maintains acceptable sanitation levels when stored. Condensate, residual water, and so on can result in grossly contaminated equipment by the time of use.

In-Process Materials

It is common practice to manufacture large batches of intermediates, such as coating solutions and lubricant blends. These should have microbial specifications, validated use lives, and a monitoring program.

Some solutions are used hot. However, if a temperature of 60°C or more is not continually maintained, these materials are at risk. For example, a sugar solution is manufactured on Wednesday and kept hot until Friday when the line is shut down for the weekend. Monday morning, the solution is reheated. However, an osmophilic yeast has grown in the solution. The yeast is killed by the heat, but the tablets are now being coated with a foul smelling, mycotoxin-containing, greenish slime solution.

Processing Methods and Parameters

Tablet cores are made in several different ways. Two common methods are direct compression of a dry blend or wet granulation. The bioburden of a dry blend is normally the direct result of the raw materials and equipment. Wet granulations are mixed with either alcohol or water and then dried. A common miscalculation is that alcohol will reliably kill contaminating microorganisms. This effect is not true unless validated. Water granulations are very vulnerable to developing high levels of microorganisms, particularly if stored before further processing.

In drying a granulation blend, the assumption is made that the heat involved will kill any contaminants—a major miscalculation. Granulations are dried to a particular moisture level or dryer exit temperature. It cannot be assumed that all of the granulation will be held at a particular temperature for a minimum amount of time. A validated time-temperature relationship is critical if it will be used to predict bioburden reduction. Normal dryer temperatures will not kill bacterial spores that are heat and chemical (alcohol) resistant. Another factor that is common to all processes is environmental contamination from air handling systems, vents, and dust collectors.

A persistent myth in the pharmaceutical industry is that a tablet press will generate sufficient pressure and resulting heat of compression to kill any microbial contaminants. Unfortunately, validation efforts show this is a very unreliable control method.

Storage Methods

Tablets are hygroscopic, including many coated tablets. Consequently, the manufacturer needs to pay particular attention to humidity and other environmental factors during tablet storage.

Packaging

The package design should provide a reliable moisture barrier, particularly under anticipated conditions of consumer use. Packaging operations should be designed to minimize contamination from equipment and the environment.

Subsequent Batch Surveillance

Many things can happen, and much remains to be done, after a product has been validated and released for sale to the general public. The QC group must be continuously involved, in addition to the routine testing described in this chapter.

Change control procedures should be in place and reviewed on a regular basis. Stability studies should be updated and continually reviewed for confirmation and possible extension of expiration dates. Failure investigations are important to track and take action where necessary. Records should be reviewed for any deviations that may have occurred during any part of the manufacturing process and then used as a basis for making improvements.

Plans should be in place to review any product returned to the company. Consumer complaints are the report card for products. Specific programs should be in place to disseminate the results of the regulatory-required investigation of each complaint by key personnel from R&D, marketing, manufacturing, and QC.

Annual reports should be thoroughly reviewed by appropriate personnel. The data and information gathered from the subsequent batch surveillance program should be carefully evaluated and applied. Process optimization can sometimes be accomplished using information collected for annual reports, as well as specification adjustments (usually tightening). New market and product opportunities also may arise from the information gathered. Purchasing opportunities may present themselves as production increases.

Opportunities for improvement in pharmaceutical manufacturing will always be available for the inquisitive ones who have gone that extra mile to learn the intimacies of the process and who seek continuous improvement. To them will go the greatest satisfaction and rewards.

REFERENCES

Cartensen, J. T., and E. Nelson. 1976. Terminology regarding labelled and contained amounts in dosage forms. *J. Pharm. Sci.* 65 (2):311–312.

Davis, J. 1978. The dating game. Paper presented at The Proprietary Association's Manufacturing Controls Seminar in Cherry Hill, New Jersey.

FDA. 1987. *Guideline for submitting documentation for the stability of human drugs and biologics.* Rockville, Md., USA: Food and Drug Administration, Center for Drugs and Biologics.

FDA. 1996. *Certain requirements for finished pharmaceuticals.* Proposed Rule, Docket No. 95N-0362.

FR. 1994. International Conference on Harmonization of Technical Requirements for Registration of Pharmaceuticals for Human Use. *Federal Register* 59 (183):48754–48759.

21 CFR 211. Current Good Manufacturing Practices.

USP 23/NF 18. 1995. *U.S. Pharmacopeia/National Formulary.* Rockville, Md., USA: U.S. Pharmacopeial Convention, Inc.

BIBLIOGRAPHY

Addison, J. F. 1981. Organization of the quality control unit. Paper presented at The Proprietary Association's Manufacturing Controls Seminar in Philadelphia.

Addison, J. F. 1985. A total quality control computer information system. Paper presented at the American Society of Quality Control Conference in Los Angeles.

Addison, J. F. 1991. Label control from industry's perspective. Paper presented at PharmTech Conference in New Brunswick, New Jersey.

Addison, J. F. 1996. Validation: Good business sense. A how-to seminar. Joint FDA/NDMA Seminar Presented in North Carolina, Puerto Rico, Colorado, and New Jersey.

FDA. 1993. *Guide to inspection of pharmaceutical quality control laboratories.* Rockville, Md., USA: Food and Drug Administration, The Division of Field Investigations.

FDA. 1994. *Guide to inspections of oral solid dosage forms: Pre/post approval issues for development and validation.* Rockville, Md., USA: Food and Drug Administration.

The Gold Sheet. 1997. 31 (1).

INDEX

Drug Manufacturing Technology Series

KEY CONCEPT CROSS–REFERENCE INDEX FOR NONSTERILE DOSAGE FORMS

Note: This key concept index is designed to reflect all of the major topics of the volumes of the *Drug Manufacturing Technology Series* that relate to Nonsterile Dosage Forms as they are published. It will be expanded for each volume of the Nonsterile group. Volume numbers are identified by an Arabic numeral followed by a colon.

9 780367 400330